US Self-Propelled Guns in action

By Jim Mesko

Color by Don Greer

Illustrated by Richard Hudson

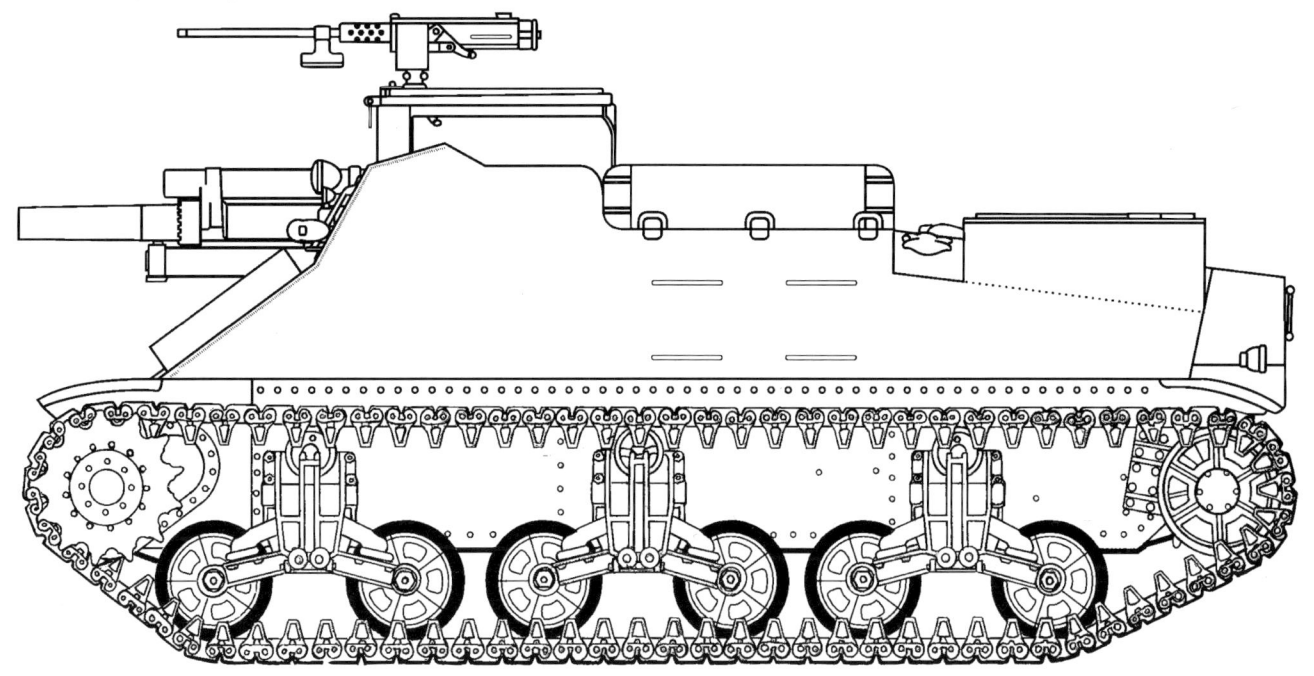

Armor Number 38

squadron/signal publications

(Cover) A late production M7 Howitzer Motor Carriage fires in support of a US attack near Kleinblitterdorf, Germany on 7 December 1944. CRAZY HELEN/ALL AMERICAN was assigned to Battery A, 231st Armored Field Artillery Battalion, 6th Armored Division.

Dedication:

To B.J. "Second Star To The Right, Then On Until Morning". ENJOY RETIREMENT!

Acknowledgements:

As is normal for such a book, numerous people contributed to what appears here. The staff of the Patton Museum proved exceedingly helpful in locating material, photographs, and for allowing access to the Robert J. Icks Collection. The National Archives staff also proved very helpful in locating obscure material. A special thanks to Jacque Littlefield who gave me the free run of his magnificent armor collection to photograph and Mike Green for his help and camera support. Richard Hunnicutt contributed much in the way of material and background and the author owes this great gentleman a debt of gratitude for all his support over the years. Blair Yoshida provided valuable information and photographs on the Sexton. Finally, a special thanks to my lovely wife Pat who spent an entire day with me carrying camera gear and suffering through countless numbers of tanks one very hot July day in California.

National Archives (NA)
US Army
US Marine Corps
ECPA
Canadian National Archives
George Balin
Blair Yoshida
Patton Armor Museum and Robert J. Icks Collection (PAM)
Don Greer
Alan Godec

COPYRIGHT 1999 SQUADRON/SIGNAL PUBLICATIONS, INC.
1115 CROWLEY DRIVE CARROLLTON, TEXAS 75011-5010
All rights reserved. No part of this publication may be reproduced, stored in a retrieval system or transmitted in any form by means electrical, mechanical or otherwise, without written permission of the publisher.

ISBN 0-89747-403-1

If you have any photographs of aircraft, armor, soldiers or ships of any nation, particularly wartime snapshots, why not share them with us and help make Squadron/Signal's books all the more interesting and complete in the future. Any photograph sent to us will be copied and the original returned. The donor will be fully credited for any photos used. Please send them to:

Squadron/Signal Publications, Inc.
1115 Crowley Drive
Carrollton, TX 75011-5010

Если у вас есть фотографии самолетов, вооружения, солдат или кораблей любой страны, особенно, снимки времён войны, поделитесь с нами и помогите сделать новые книги издательства Эскадрон/Сигнал еще интереснее. Мы переснимем ваши фотографии и вернём оригиналы. Имена приславших снимки будут сопровождать все опубликованные фотографии. Пожалуйста, присылайте фотографии по адресу:

Squadron/Signal Publications, Inc.
1115 Crowley Drive
Carrollton, TX 75011-5010

軍用機、装甲車両、兵士、軍艦などの写真を 所持しておられる方はいらっしゃいませんか？どの国のものでも結構です。作戦中に撮影されたものが特に良いのです。Squadron/Signal社の出版する刊行物において、このような写真は内容を一層充実し、興味深くすることができます。当方にお送り頂いた写真は、複写の後お返しいたします。出版物中に写真を使用した場合は、必ず提供者のお名前を明記させて頂きます。お写真は下記にご送付ください。

Squadron/Signal Publications, Inc.
1115 Crowley Drive
Carrollton, TX 75011-5010

(Right) An early production M7 from the 2nd Armored Division moves past a crowd of curious Italian civilians on the outskirts of Sciacca, Sicily on 20 July 1943. This early production M7 features the three piece transmission housing and the vertical volute suspension with the return roller mounted above the center of the bogie assembly. The circle around the star was first used during OPERATION HUSKY, the invasion of Sicily, in order to avoid confusion with the German cross at a distance. The star on the upper half of the transmission housing is painted on a blue disk. This insignia was used during the later stages of the North African campaign. A blue circle separated the white star from the sand colored paint scheme used on some vehicles. (USA/NA)

Introduction

Artillery is the unsung combat arm of the battlefield. From its earliest inception to today, artillery has rarely received the coverage given the other, more glamorous combat arms such as infantry, airborne, or armor. Many battlefield successes might never have occurred without the use of artillery — the timely arrival of an incoming artillery barrage often spelled the difference between victory and defeat.

Despite its importance on the battlefield, artillery was often handicapped by its lack of mobility and the time needed to set up firing positions. This lack of mobility was particularly evident during World War One due to the increasing size and complexity of the artillery pieces which had become the backbone of the armies on both sides of the trench line. In the event of a successful breakthrough, horse drawn artillery could not be quickly moved forward and set up due to the scarred battlefield terrain. Additionally, terrain often hindered the movement of ammunition supplies used to both further support a breakthrough and prevent an enemy counterattack.

With the advent of the tank, some work began on mounting artillery weapons on tractor chassis, but World War One ended before much progress was made. Nevertheless, the seeds had been sown for the future development of a new type of artillery piece which was able to move under its own power and keep pace with changing battlefield conditions.

Simply called 'self-propelled artillery' or 'SP's' these weapons would eventually evolve into a deadly partnership with infantry and armor to form a combined arms team which today dominates the battlefield. This deadly partnership was amply demonstrated during OPERATION DESERT STORM in the late winter of 1991.

Following World War One the major military powers carried out experiments with mechanized forces and developed a variety of prototype self-propelled artillery mounts. Although some of these self-propelled mounts showed promise, there was no widespread production of these pieces due to the conservative nature of most armies during the 1920s and 1930s. Then, in September of 1939, German dictator Adolf Hitler initiated World War Two with a lightning attack, known as the *Blitzkrieg*, on Poland. This campaign saw the introduction of highly mobile German mechanized and armored units. These units quickly cut through the courageous, but hopelessly antiquated Polish Army. Poland fell by the end of September.

The following spring it was the turn of Western Europe to feel the crunch of the German Blitzkrieg, and by the summer of 1940, only England defiantly held out against Hitler. Part of the success of the Blitzkrieg was due to the use of 'flying artillery' — the infamous Ju-87 Stuka dive bomber. Additionally, the Germans had modified some obsolete light tank chassis to carry artillery pieces. This ability to provide immediate fire support to their fast moving armored columns had allowed the Germans to quickly deal with points of resistance and move on as well as provide fire support during vital river crossings. The Allied command structure, still thinking of the static nature of World War One, was thrown into great confusion by the speed of the German advance. This confusion, and the Allies' inability to counter the German attacks, greatly contributed to the stunning German victory.

American political and military leaders viewed Germany's success with great anxiety. During 1939 and 1940 the US military was in a sad state of affairs as a result of years of neglect and financial restrictions placed on the development of new weapons. Some work had been carried out on the development of a new armored force which included some experimentation with self-propelled artillery, but in 1940 there was little to show for these experiments. Additionally, the US Army was led for years by conservative generals. The effects of the Blitzkrieg caused a great shakeup within the US Army hierarchy and awoke many people within the US government. Efforts were soon underway to build up US military forces. Since there was a great need to quickly build up the US Army's mechanized forces, it was necessary to modify existing equipment to serve as an interim solution until more advanced weapons could be developed and produced.

The M2/M3 Half-Track was coming into service in 1940, and since it was capable of off-road operation, the US Army decided to fit the M3 with the standard 75mm M1897A5 field

The M3 Half-Track was used for a number of early self-propelled gun mounts. The first vehicle, the 75mm M3 Gun Motor Carriage, was designed as a tank destroyer, however, the M3 was used to provide fire support in the Philippines rather than in its intended role. The M3 reverted to the tank destroyer role in North Africa, but was later used by the US Marines and the British Army for both direct and indirect fire support. (PAM)

The T30 was the next self-propelled gun version of the M3. The vehicle mounted a 75mm M1A1 pack howitzer in a box-like shield. The T30 saw service in North Africa, Sicily, and Italy until replaced by the M8 Howitzer Motor Carriage. (PAM)

gun. This gun was an American version of the famous 'French 75' from World War One. The combination of the M3 Half-Track and the M1897A5 gun was initially designated the T12. After successful trials the T12 was accepted into service as the 75mm M3 Gun Motor Carriage (GMC). The M3 GMC was originally intended as a mobile anti-tank vehicle, however, the vehicle saw its first combat use in the Philippines as a mobile artillery unit supporting the 192nd and 194th Tank Battalions during the retreat to the Bataan peninsula. M3 Gun Motor Carriages were later used in both the anti-tank and self-propelled artillery roles in North Africa, Sicily, and the Pacific until replaced by newer vehicles or retired by attrition.

Two additional versions of the M3 Half-Track were developed in response to an Armored Force requirement for an assault gun to provide support for tank and armored reconnaissance units. The first version mounted a 75mm M1A1 pack howitzer and received the designation T30 Howitzer Motor Carriage (HMC). The second version mounted a 105mm M2A1 howitzer under the designation T19 HMC. Neither the T30 or the T19 HMCs received a production (M) designation since they were viewed as interim solutions until they could be replaced by purpose-built vehicles coming off the assembly line.*

The weight and recoil of the 75mm howitzer did not overly stress the M3 Half-Track chassis, however, the 105mm howitzer pushed the vehicle's structure to its limits. Despite these concerns, both the T30 Howitzer Motor Carriage and the T19 HMC saw combat in North Africa, Sicily and the early stages of the Italian campaign before being retired. Although their performance was acceptable considering the ad hoc nature of their development, both the T30 and the T19 underlined the need for purpose built self-propelled vehicles which offered better protection and mobility.**

During the fall of 1941, while work progressed on half-track mounted artillery pieces, the US Army began to investigate mounting a 105mm howitzer on a tank chassis to increase fire

*The T designation was given to prototype and test vehicles and to vehicles approved for limited production. Vehicles approved for full-scale production as standard issue were given an M designation.
**For additional information on these vehicles see M3 Half-Track in Action, Number 34, by Squadron/Signal Publications.

support for the newly emerging armored divisions. The chassis of the M3 Lee medium tank, which had just entered full scale production, was selected as the basis for a new self-propelled artillery mount. Two M3 chassis were manufactured by the Baldwin Locomotive Works and equipped with a standard 105mm M2A1 howitzer mounted slightly to the right of center in a box-like open topped structure. Both vehicles were designated the 105mm T32 Howitzer Motor Carriage (HMC) and shipped to Aberdeen Proving Ground, Maryland for initial trials. One T32 HMC vehicle was then sent from Aberdeen to Fort Knox, Kentucky for further evaluation by the Armored Forces Board. This vehicle was subjected to an extensive series of tests over a three-day period. The board was impressed with what they saw, but recommended a few changes to the T32's basic design.

These recommendations included reducing the superstructure armor from 3/4 to 1/2 inch, raising the front armor plate by three inches while lowering the side and rear plates by 11 inches, and increasing the total traverse of the howitzer from 31° to 45°. The Armored Forces Board also accepted a reduction in the 105mm howitzer's elevation from the specified +65° to +35° in order to maintain a low vehicle silhouette. Further changes included the addition of a .50 caliber anti-aircraft machine gun mount in the forward starboard corner of the fighting compartment along with a folding machine gun mount on the rear deck. Additionally, a transmission access plate was added and the internal ammunition storage was increased from 44 to 57 rounds.

Work began at Aberdeen Proving Ground during February of 1942 incorporating these changes to the number two pilot model. The fighting compartment of the redesigned T32 Howitzer Motor Carriage differed greatly from the original pilot model's appearance. The front armor was reshaped to provide more interior room and allow the mounting of the anti-aircraft machine gun ring. Internal changes for the 105mm howitzer and ammunition storage were also made. This modified T32 was then shipped to the American Locomotive Company for use as a production prototype. The new vehicle was redesignated the 105mm M7 Howitzer Motor Carriage in April of 1942.

The T19 was the final self-propelled artillery variant of the M3 Half-Track. The vehicle carried a 105mm M2A1 howitzer. The weight and recoil of the howitzer, the largest weapon carried by the M3, undoubtedly taxed the chassis to its limits. The T19 saw brief service in North Africa, but was quickly replaced by the superior M7 Priest which utilized the M3 Lee medium tank chassis. (PAM)

The first fully tracked US self-propelled gun was the T32. The T32 was based on the M3 medium tank chassis and mounted a 105mm M2A1 howitzer in an open topped fighting compartment. Two pilot models were ordered during October of 1941. The first pilot model was tested in early February of 1942. Only a few minor modifications were needed prior to placing the T32 into production under the designation M7 Howitzer Motor Carriage. (PAM)

Development

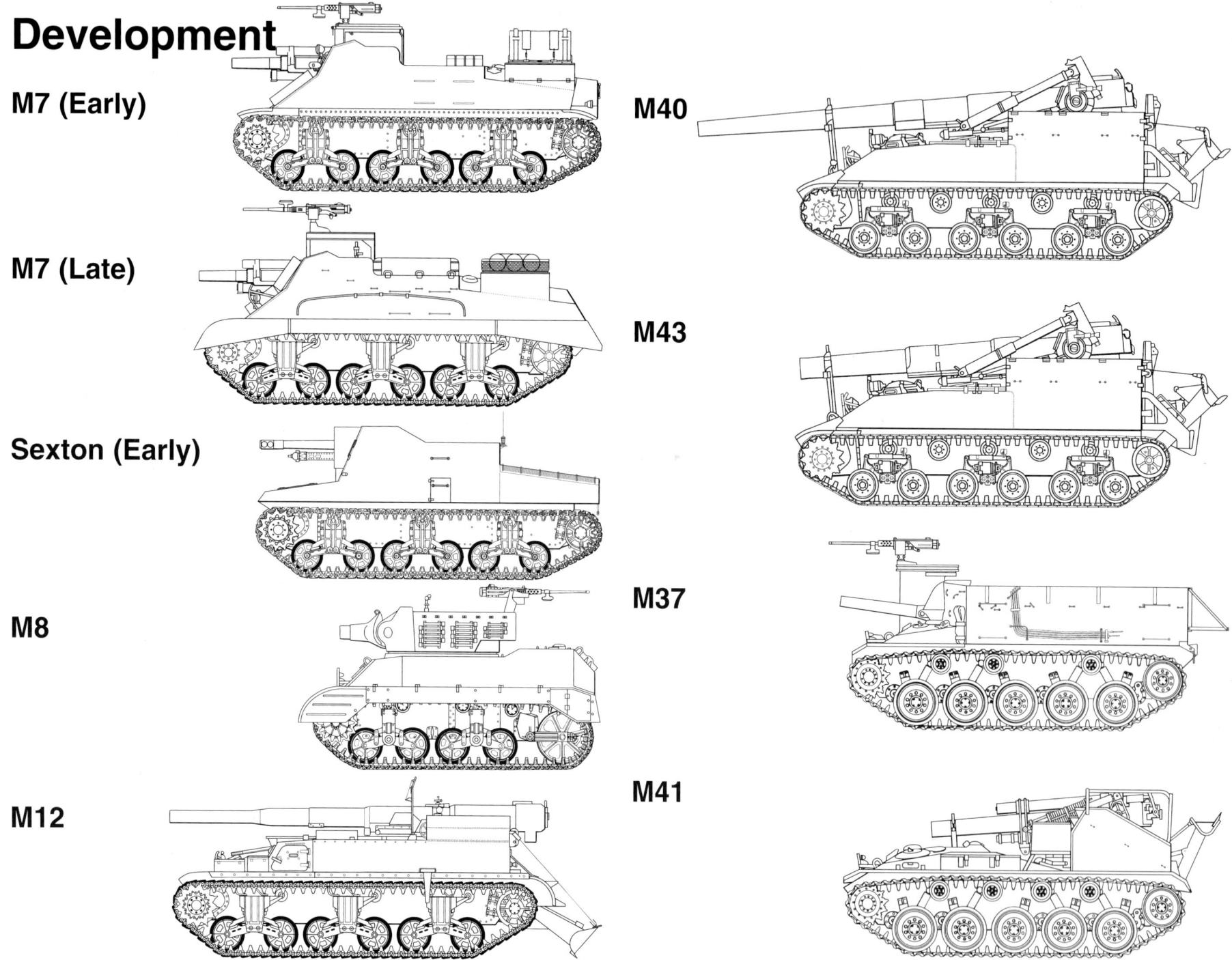

M7 (Early)

M7 (Late)

Sexton (Early)

M8

M12

M40

M43

M37

M41

M7 Priest Howitzer Motor Carriage

The M7 Howitzer Motor Carriage (HMC) was standardized for production in April of 1942. The M7 used the standard M3 Lee medium tank chassis with a large, open topped structure replacing the armored fighting compartment and turret.

The M7 was equipped with a 105mm M2A1 howitzer fitted slightly to the right of the vehicle's centerline. The howitzer had a 45° traverse — 15° to port and 30° to starboard — along with -5° to +35° of elevation. The initial model of the M7 could carry 57 rounds, however, this was later increased to 69 rounds by reducing the number of folding crew seats in the fighting compartment. The 105mm howitzer fired a full range of ammunition types, including high explosive (HE), high explosive anti-tank (HEAT), white phosphorous (WP), and smoke. Maximum range of the howitzer was 11,400 yards (10,424 meters) with a sustained rate of fire of eight rounds per minute.

A .50 caliber M2 anti-aircraft machine gun ring mount was fitted into the upper starboard corner of the fighting compartment. This mount protruded slightly outside of the fighting compartment and was protected by a curved strip of armor plate. Additionally, a small ledge joined the bottom of the mount and the hull. On later models the armor plate was extended further down the hull and the ledge served as a seat for the machine gunner. This machine gun position led to M7's nickname — Priest. According to legend, when British troops first received the vehicle in North Africa, the machine gun mount reminded them of a church pulpit.

The driver sat in the lower port side of the M7 HMC. The driver was provided with a large hatch for direct vision when open. The hatch incorporated a protectoscope for indirect vision when the hatch was closed. A removable windshield was also provided to help keep out dirt, dust, and rain when the driver's hatch was open. A foul-weather canvas tarp was used to cover the open fighting compartment during periods of inclement weather. The M7 was manned by a crew of six: commander, driver, gunner, and three ammunition handlers.

Early production M7s used the early three piece differential housing from the M3 medium tank. This differential housing featured a small cutout in the upper starboard corner that provided clearance for the sponson mounted 75mm gun on the M3. Later production vehicles used the three piece housing from an M4 Sherman medium tank which lacked the cutout. Late production M7s used a one piece casting. The one piece differential housing was produced in two versions — initially a rounded version, and finally a sharply pointed version.

Initially the M7 used the vertical volute suspension (VVS) system found on the M3 medium tank which featured the track return roller centered above the bogie assembly. A heavier duty bogie assembly which had the return roller mounted at the rear was developed later and put into production during the summer of 1942. At first this suspension was used solely on the M4 which had first priority since it was heavier than the M3. Late model M7s received the later heavy duty bogie assembly when supplies eventually met demand.

The M7 HMC was initially equipped with storage boxes on both sides of the rear deck. These boxes had side hinging doors to allow the mounting of external gasoline tanks on top of the boxes. These tanks were equipped with quick release mounts which allowed the tanks to be jettisoned from inside the fighting compartment. This system, however, was deleted early in the production run. The storage boxes were redesigned with top hinging hatches. Wire mesh baskets were fitted on top of the boxes for additional storage.

Power was supplied by a Continental R975 air-cooled radial engine which provided a speed of up to 24 miles (38.6 km) per hour. The M7 had a cruising range of approximately 120 miles (193 km) on 175 gallons (662.4 liters) of gasoline. The vehicle weighed 50,700 pounds (22,997 kg) fully loaded.

Further testing and reports from the battlefield, along with improvements in M4 components, led to additional changes on the M7 Howitzer Motor Carriage. One of the earliest battlefield reports indicated that the tips of the vertically stowed ammunition were exposed to enemy fire. Some units began to add armor plates to the upper hull sides to protect the ammunition. Hinged armor flaps were then developed at the factory and installed on the sides and rear of late production M7s. Armor flap kits were also developed for crews to retrofit existing vehicles in the field. American Locomotive manufactured 500 improved M7s with a further 176 produced by the Federal Machine and Welder Company. Combined with the initial run of 2814 vehicles by American Locomotive, a total of 3490 M7 HMCs were manufactured when production ended in the spring of 1945.

M7B1

The M7 HMC was originally based on the chassis of the now obsolete M3 Lee medium tank. Later production M7s began to employ components of the M4 Sherman medium tank. Late production M7s used the chassis of the M4A3 Sherman medium tank in lieu of the earlier M3 Lee chassis. M7s based on the M4A3 chassis were designated M7B1. Both the M7B1 HMC and the M4A3 medium tank were powered by an eight cylinder, liquid-cooled Ford GAA engine. The Ford GAA engine provided a maximum speed of 26 miles (41.9 km) per hour — two miles (3.2 km) per hour faster than the M7. Cruising range for the M7B1 was the same as for the M7, approximately 120 miles (193 km) on a fuel capacity of 175 gallons (661.5 liters).

M7B1s also used the vertical volute suspension (VVS) system with heavy duty bogie assemblies found on the M4A3 tank and late production M7 HMCs. The M7B1 retained the armament of the M7 — one 105mm M2A1 howitzer with 69 rounds of ammunition and one .50 caliber M2 anti-aircraft machine gun. Aside from the engine and rear deck the M7B1 was similar

The first M7 Howitzer Motor Carriage came off the assembly line during April of 1942. Early production M7s featured a three piece transmission housing and the vertical volute suspension with the return roller centered above the bogie assembly. The empty boxes on the hull front were used to hold track grousers. (PAM)

The M7 had a machine gun mount on the right side of the fighting compartment. Early M7s had a short armored ring around the mount, however, this was later extended down the side. This mount led to British troops in North Africa giving the M7 the nickname Priest, due to its resemblance to a church pulpit. This vehicle has folding armor panels behind the pulpit and at the rear of the fighting compartment to protect the vertically stored ammunition whose tips were previously exposed above the side armor. (PAM)

The ring on top of the pulpit provided a 360° field of fire for the .50 caliber M2 machine gun. The object on the right side of the pulpit interior is the bracket for the gunner's seat. (Author)

in appearance to the late production M7. The Pressed Steel Car Company manufactured 826 M7B1 HMCs from March of 1944 to February of 1945.

M7B2

Production of the M7 and M7B1 Howitzer Motor Carriages ended in 1945, however, there was one further model which was to see service. During the Korean War troops found the 105mm howitzer's +35° elevation was insufficient for placing rounds on the reverse slopes of Korea's high hills and mountains. The +35° elevation was designed to prevent the weapon from recoiling into the floor plates of the fighting compartment. The same situation had occurred in Italy during World War Two, however, the solution had been to drive the M7s onto steep earthen ramps in order to increase the weapon's elevation. Although this method was initially used in Korea, a better solution was needed. To increase the howitzer's elevation, the mount was raised to provide the additional recoil clearance and allow the howitzer to be raised

The 105mm howitzer was mounted slightly to the right of the vehicle centerline in the open topped fighting compartment. The driver sat to the left of the gun behind the armored flap. Vertical storage racks for the ammunition were located on both sides of the fighting compartment. Curved brackets for the jettisonable gasoline tanks are mounted on top of the rear storage boxes. (PAM)

Machine Gun Pulpit

Early Pulpit

Late Pulpit

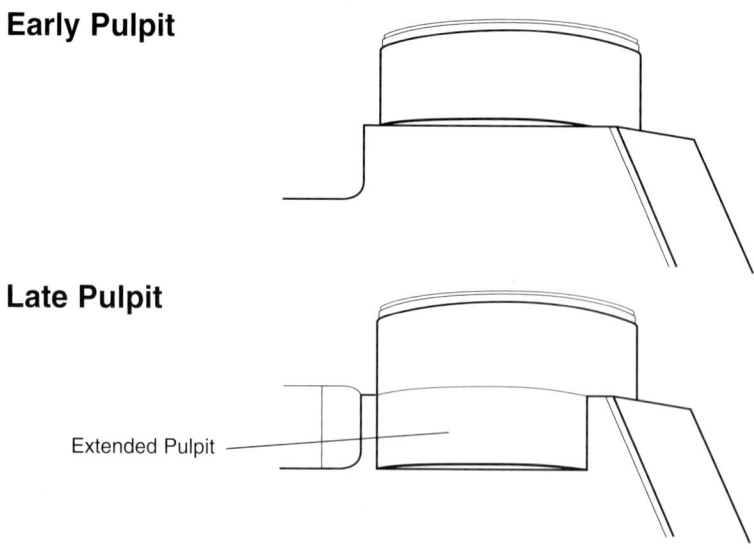

Extended Pulpit

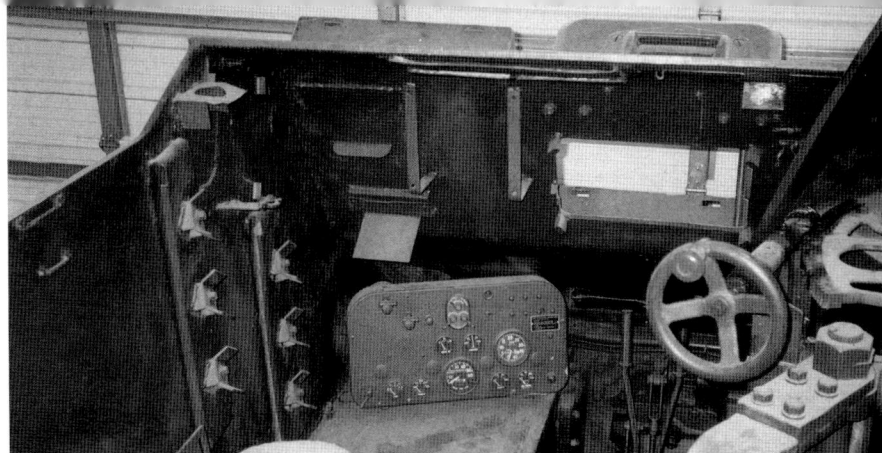

(Above) The instrument panel was located to the left of the driver. The open driver's hatch could be fitted with a windshield to keep out dust, dirt, and rain. Crew weapons were stored in the rack to the left of the instrument panel. This particular M7 is part of the fabulous collection of Jacque Littlefield. (Author)

(Right) Later production M7s featured revised vertical ammunition storage racks. The 105mm gun mount was set into the fighting compartment minus its axle and wheels. The driver's position and control panel are just to the left of the gun mount (this vehicle features a revised instrument panel). The storage boxes of later production M7s were not fitted with jettisonable gasoline tank mounts. (PAM)

to the originally specified +65°. The raised howitzer mount also required the raising of the .50 caliber machine gun mount to maintain a 360° field of fire. While these modifications raised the M7's overall height, they were considered a good trade-off for improved fire on reverse slopes. These modified vehicles, all based on the earlier M7B1, were designated M7B2 HMC. The 127 M7B2s were the last variants to remain on active duty with the US Army.

(Below) The M7 was continually revised throughout its manufacturing life. The older three piece transmission cover has been replaced with a one piece sharp-nosed casting. The vehicle has the later heavy duty vertical volute suspension with the return roller mounted on an arm trailing to the rear of the bogie assembly. This M7 also features side hinged armor panels and new rear storage boxes with top hinged lids and wire storage baskets. (PAM)

(Above) The early production M7 carried 24 rounds of ready ammunition in the fighting compartment storage racks. A removable panel at the rear of the fighting compartment provided access to the engine bay. The breech of the 105mm howitzer is open. (US Army/NA)

(Below) The 105mm M2A1 howitzer was mounted on a platform in the front of the fighting compartment. The weapon was protected by a small armored shield which moved with the howitzer. The twin I-beams transferred the gun's recoil forces to the hull of the vehicle. The gunner sat to the left of the breech assembly. (Don Greer)

(Above) Late production M7s featured revised ammunition racks with horizontal stowage. This is believed to have been a field modification designed to both carry more ammunition and improve its protection. Aside from the rack changes, both vehicles are basically the same, although they are separated by nearly three years. (PAM)

Combat Service

The M7 HMC initially saw combat with British forces in North Africa during the fighting at El Alamein during the summer of 1942. M7s first saw action with American forces when they landed in North Africa during OPERATION TORCH in November of 1942. M7 and M7B1 HMCs then saw continuous combat throughout World War Two while assigned to the field artillery battalions supporting Allied armored units. The M7s served in Sicily, Italy, northwest Europe and to a lesser degree in the Pacific during the latter stages of World War Two. In British service the M7 Priest was used by field artillery regiments until the introduction of the Sexton Self-Propelled Gun. The Priest was then put to other uses, the most notable of which was the Kangaroo Armored Personnel Carrier (APC). The Kangaroo APC, also known as a 'defrocked Priest', had the 105mm howitzer and the ammunition storage bins removed and the resulting aperture in the front armor plated over. The Kangaroos proved popular for moving troops under fire due to its armored protection and mobility over difficult terrain.

After World War Two many M7 HMCs were either scrapped or supplied to American allies under various defense aid agreements. Other M7 and M7B1s remained in US Army service throughout the 1940s and into the 1950s. The M7 last saw action with American forces during the Korean War. Other nations also used the M7 in various conflicts, including Pakistan during its 1965 and 1971 wars with India, and Israel during the 1967 Six-Day War and the 1973 Yom Kippur War. The Yom Kippur War is believed to be the last major use of the M7 HMC in combat. The few vehicles which remain are now either awaiting scrap, serving as monuments, or in the hands of collectors and museums.

Later production M7s used the M4A3 Sherman tank chassis. These vehicles were designated M7B1, and were identical to the earlier M7 with the exception of the engine, engine deck, and rear hull. The wider engine grill covered nearly the entire top of the deck. The wire baskets on top of the rear storage boxes were typical of later production M7s. (Author)

During the Korean War the need to increase the elevation of the 105mm howitzer resulted in the M7B2. This model had the gun mount raised to increase the gun elevation to a full 65°. The machine gun mount was also increased in height to maintain a clear 360° field of fire. (PAM)

Suspension

Early VVSS

Later VVSS

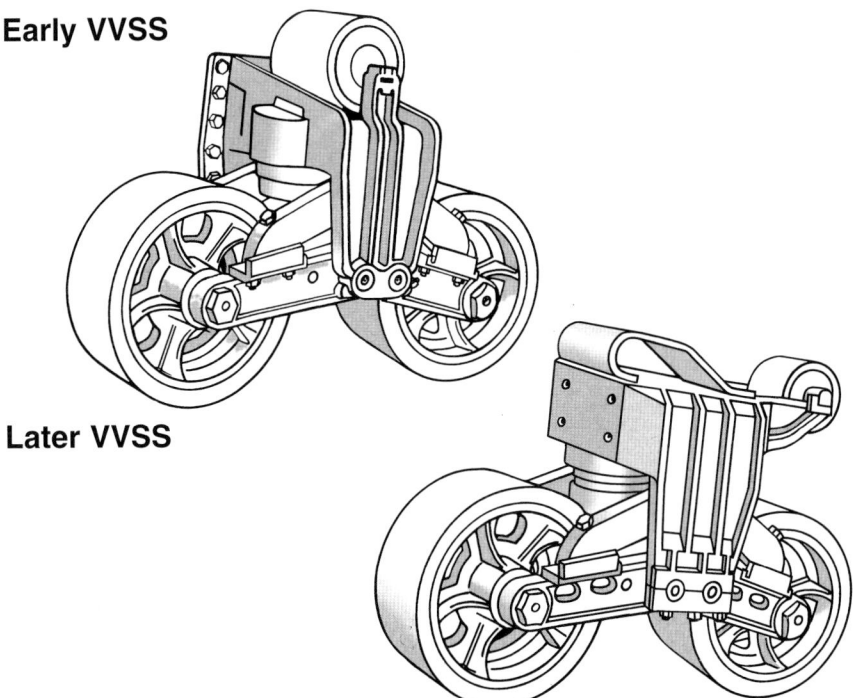

The M7 first saw action in North Africa with British forces at El Alamein. These American troops were part of a technical team which was sent with the M7s to show the British how to operate and maintain the vehicles. This experience also gave the Americans valuable desert training and allowed them to observe the M7 under combat conditions. The early production Priest in the background is equipped with the short .50 caliber machine gun pulpit. (PAM)

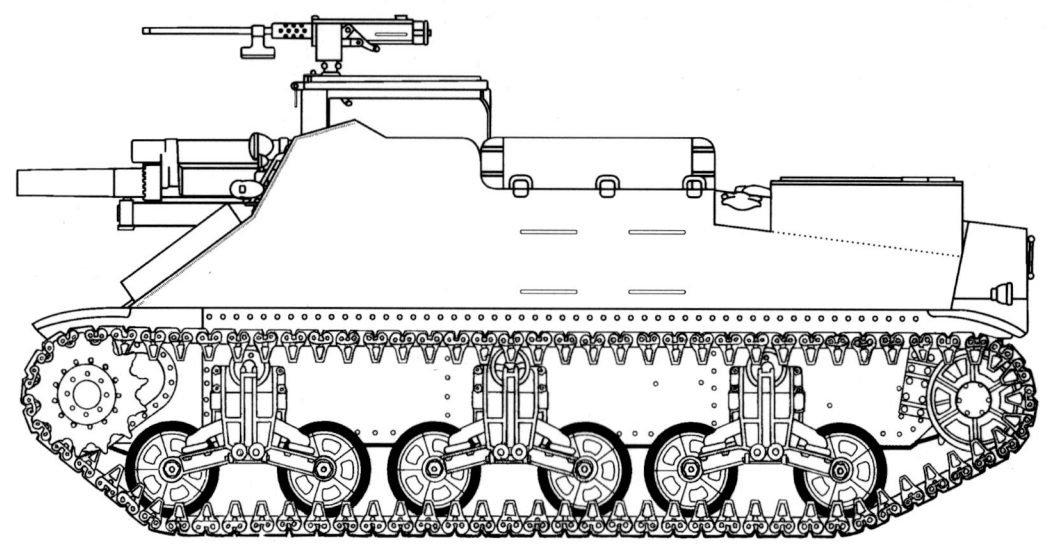
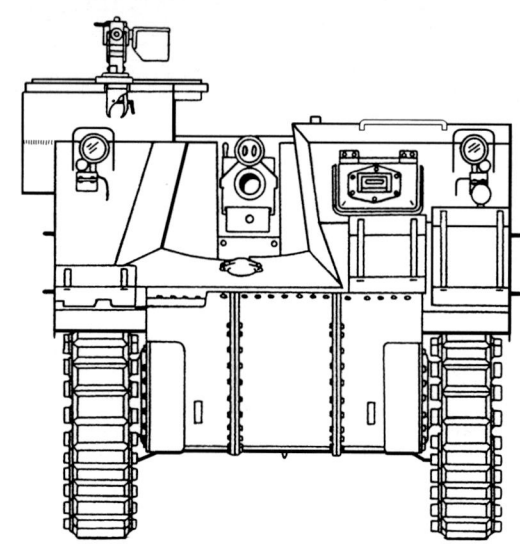

M7 Priest HMC Specifications

Length...............19 Feet, 9 Inches
Width...................9 Feet, 5 1/16 Inches
Height..................9 Feet, 6 Inches (w/machine gun)
Weight................50,600 Pounds
Armament..........1 x 105mm M2A1 howitzer, 1 x .50 cal M2 machine gun.
Speed.................24 MPH
Crew...................6

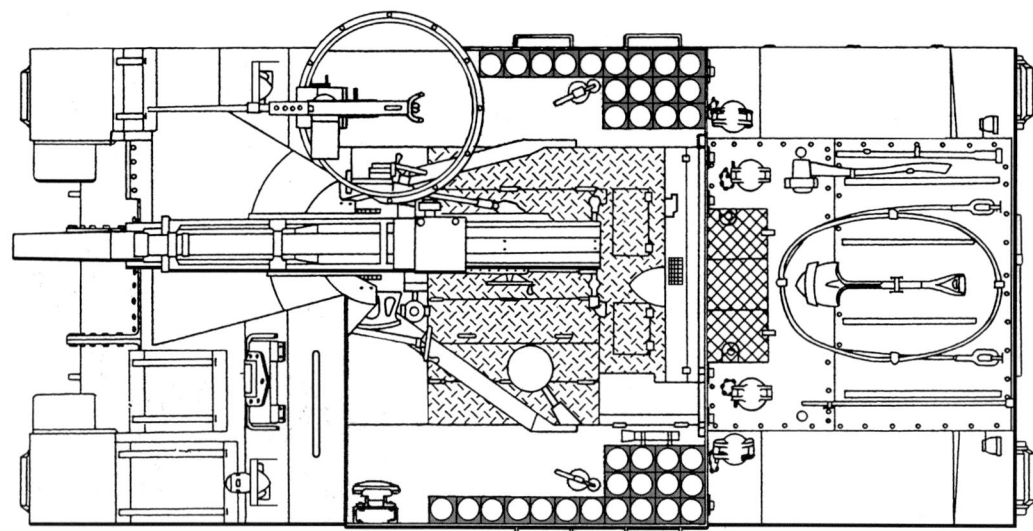

(Above) An early production M7 rolls through Sciacca, Sicily on 20 July 1943. Track grousers are stowed in boxes on the glacis plate, while a camouflage net is tied down on the fender in front of the machine gun pulpit. The M7 is towing an M8 ammunition trailer. (US Army/NA)

(Above Right) Following its combat debut in North Africa, the M7 was next used in OPERATION HUSKY — the invasion of Sicily during July of 1943. This heavily laden M7 moves through the town of Menfi shortly after the invasion. Aside from the large amount of extra gear, this vehicle has been fitted with a rack of land mines on its side, an uncommon modification for an M7. (US Army/NA)

(Right) The French received Lend-Lease M7s to help rebuild the Free French Army under the command of General Charles de Gaulle. These M7s from the 3rd Algerian Infantry Division sit at an intersection in Marseilles after the city's liberation during the summer of 1944. (US Army/NA)

The seven-man crew of LE VENGEVER pose by their M7 during the latter half of 1944. This early production M7 was assigned to 1st Battery, 3rd Regiment of Colonial Artillery, French 2nd Armored Division. The vehicle lacks the side armor extensions to protect the vertically stowed ammunition in the fighting compartment. (Alan Godec)

A well stowed early production M7 fires on German positions in the Rhine River Valley on 9 December 1944. This M7 has the T49 steel parallel bar track and has been fitted with extra racks for jerry cans over the rear fenders. The side and rear armor extensions are welded in place. A pioneer rack, another field modification, has been fitted under the short machine gun pulpit. (US Army/NA)

The crew of Mable has just fired the 105mm howitzer and the barrel is in the full recoil position. Mable is equipped with the early vertical volute suspension, which is believed to have been painted a lighter color to help break up the contrast between the hull and the suspension. The M7's lower hull was sometimes painted white to lessen the effect of shadows caused by the overhanging upper hull. (US Army/NA)

M8 Ammunition Trailer

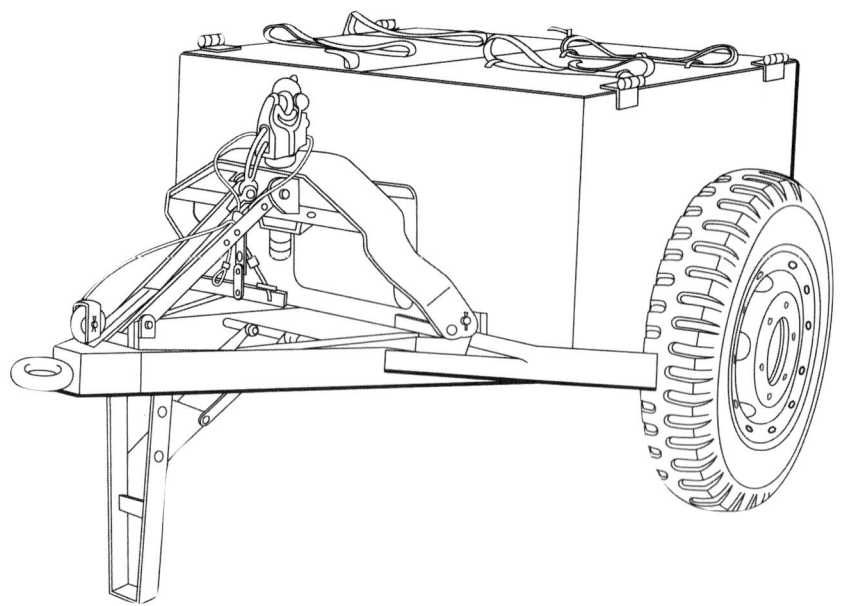

(Below) French soldiers clean the 105mm howitzer barrel on their M7 in late 1944. This vehicle was assigned to 2nd Battery, 3rd Regiment of Colonial Artillery, French 2nd Armored Division. The M7 is equipped with the heavy duty vertical volute spring suspension. An M8 ammunition trailer is parked behind the M7. (Alan Godec)

(Above) This M7 from the 5th French Armored Division has been busy firing on German positions near Keysersberg, France on 17 December 1944. The howitzer rounds were shipped in cardboard packing tubes to protect them from the elements. The white diamond on the side is surrounded by alternating red and blue triangular blocks. (US Army/NA)

(Below) A well camouflaged M7 from the 212th Armored Field Artillery Battalion, 6th Armored Division sits in the snow near Bastogne, Belgium on 8 January 1945. A tarp under the netting helps keep the fighting compartment and crew dry. This vehicle is equipped with an unusual field modified storage rack on the front. The 212th AFAB fired 1000 rounds in one day — twice the normal amount — and set a battalion record. (US Army/NA)

(Above) An M7 from the 22nd Armored Field Artillery, 4th Armored Division, crosses a Bailey bridge over the Main River near Hanau, Germany on 28 March 1945. The M7 is pulling an M8 ammunition trailer. The crew has welded bars across the front of the M7 to retain extra stowage. (US Army/NA)

(Below) The M7 saw only limited use in the Pacific due to the small islands and jungle terrain. They were used mainly in the Philippines and on Okinawa. This M7 from the 637th Tank Destroyer Battalion moves past the remnants of destroyed Japanese vehicles while crossing the Magot River on Luzon. The 637th TD Battalion later converted to the M18 Hellcat tank destroyer. (US Army/NA)

(Above) This M7 from Battery A, 1125th Field Artillery Battalion, has had jerry can racks added to the rear quarters of the hull and a large rack welded to the rear of the engine deck. This M7 lacks hinged side and rear armor plates on the fighting compartment which leaves the tops of the ammunition exposed to enemy fire. (US Army/NA)

The M7 saw its last major service with US forces in Korea. Annie Oakley was attached to the 24th Infantry Division and took part in a counter-offensive against Chinese forces trying to break through UN lines during the spring of 1951. (US Army/NA)

M7 Priest/Kangaroo

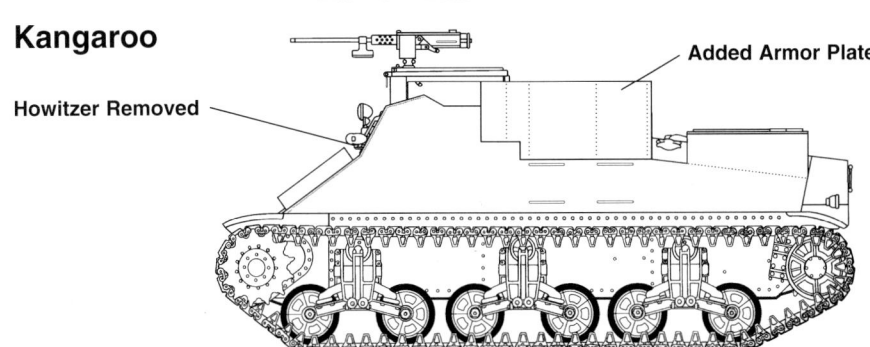

The Nationalist Chinese retreated to the island of Formosa in 1949 following their defeat by the Communists during the Chinese Civil War. The Nationalist Chinese Army was rebuilt with American military aid which included the delivery of surplus M7 HMCs. It is believed that no Nationalist Chinese M7s saw combat against the Communists. (PAM)

The introduction of the Sexton SPG led to many M7 Priests being 'defrocked' and used as Kangaroo Armored Personnel Carriers. A large armored plate was welded over the empty gun position and troop benches were installed in the fighting compartment. The Kangaroo proved popular for moving troops into combat and was able to keep pace with armored units. (Canadian Public Archives)

Sexton

Shortly after the 105mm M7 Howitzer Motor Carriage (HMC) entered production in April of 1942, the British General Staff indicated their need for a similar type of self-propelled weapon to mount the 25 pounder (88mm) Mark II gun. This British request prompted the revision of the second T32 to carry the 25 pounder gun in June of 1942. The reworked T32 was designated the 25 pounder T51 Gun Motor Carriage (GMC). The T51 was tested at Aberdeen Proving Ground, Maryland where the riveted gun cradle failed. The gun mount was revised to accept a later welded type of gun cradle, since the riveted type of cradle was no longer in production. T51 testing continued into early 1943, however, the T51 program was canceled due to the success of a parallel Canadian program which used the same gun.

The Canadian program involved mounting the 25 pounder Mk II gun on a Canadian Ram tank chassis. The Ram tank used the chassis of the US M3 Lee medium tank with a redesigned upper hull. The Ram was armed with a turret mounted 2 pounder (40mm) or 6 pounder (57mm) gun. The Canadian Department of National Defense requested the development of a 25 pounder (88mm) Mark II gun mount on the Ram chassis. This request was to meet both British and Canadian requirements for a light self-propelled artillery piece. The pilot model was ready for testing by late 1942 and, after extensive evaluation, only minor modifications were required for production. The vehicle was designated the 25 Pounder Ram Carrier and accepted for production in early 1943. The designation was later changed to the 25 Pounder Self-Propelled Gun, Sexton. The name Sexton kept up the British tradition of giving clerical names to their self-propelled artillery.

The Sexton was similar in appearance to the M7 Priest and employed an open topped fighting compartment, although there was no machine gun pulpit. A tarpaulin was provided to shelter the Sexton's six-man crew from inclement weather.

The 25 pounder gun was mounted slightly to the left of the centerline with the driver being repositioned to the starboard side of the fighting compartment. The 25 pounder gun could traverse through a 40° arc — 25° to port and 15° to starboard. Gun elevation ranged from -9° to +40°, however, in order to achieve this, the weapon's recoil was reduced from 36 to 20 inches (91.4 to 50.8 cm). The Sexton carried a total of 112 rounds of high explosive (HE), armor piercing (AP), and smoke projectiles for the 25 pounder gun. After production was underway, provisions were made for carrying a pair of .303 caliber Bren machine guns for close-in and anti-aircraft defense.

The Sexton, like the M7 HMC, reflected the ongoing changes in the evolution of the M3/M4 series. Early production Sextons had the three-piece final drive housing and the early vertical volute suspension with the return roller centered above the bogie assembly. Later production vehicles had the one-piece differential housing and heavier duty VVS bogies.

The Sexton was powered by a 484 horsepower, air cooled Continental R975 radial engine. The R975 engine provided a maximum speed of 25 miles (40 km) per hour. The Sexton's range was 180 miles (290 km). Performance of the Sexton was similar to the M7 although the Sexton weighed 7000 pounds (3175 kg) more than the Priest.

There were two versions of the Sexton produced: the Sexton I and the Sexton II. The main difference between the two was the addition of battery and generator boxes on the rear corners of the Sexton II. These boxes were absent on the 124 Sexton Is produced. Metal bars were welded to the battery and generator boxes to connect the boxes with the rear superstructure plate. Additional bars were welded between the two boxes across the rear of the hull. This arrangement created a large stowage area on the engine deck. Montreal Locomotive Works produced 424 Sextons by the end of 1943, however, continuing orders eventually raised the total number produced to 2150 when manufacturing ceased at the end of World War Two.

The Sexton gradually began replacing the M7 Priest within British and Canadian artillery

The Canadian Ram Tank employed a redesigned hull on the US M3 Lee medium tank chassis. This early Ram 1 was armed with a turret-mounted 2 pounder gun. Later Ram 1s were armed with a 6 pounder turret mounted gun and deleted the auxiliary machine gun turret. A small machine gun turret was mounted next to the driver's position in the front of the hull. (US Army)

During June of 1942 the second pilot T32 was equipped with a British 25 pounder gun and redesignated the T51 Gun Motor Carriage. The T51 never went into production due to successful development of a Canadian self-propelled gun based on the Canadian Ram tank chassis. (PAM)

(Above) The Sexton Self-Propelled Gun was based on the Ram tank chassis — itself based on the M3 Lee medium tank chassis. The 25 pounder Mk II gun was mounted slightly left of the centerline, with the driver seated on the starboard side. This contrasted with the M7 GMC, which had the 105mm howitzer slightly right of the centerline and the driver seated to port. (Author)

(Above Right) Early production Sextons did not have a machine gun mount although brackets were added for two .303 caliber Bren light machine guns beginning with the 147th production vehicle. One hundred twelve cartridges and projectiles could be stored within the vehicle. (PAM)

regiments in Italy and Northwest Europe. The Sexton proved to be an excellent vehicle and remained in service with both the British and Canadian armies until the mid-1950s. The redundant Priests were converted to other uses, most notably the Kangaroo Armored Personnel Carrier (APC) which had the gun removed and the opening plated over. The Kangaroo allowed infantry to be transported into battle in relative safety and were able to keep pace with armored units over the worst of terrain.

(Right) Auxiliary generator and battery boxes were added to the rear deck of the Sexton beginning with the 125th production vehicle. Vehicles equipped with these boxes received the designation Sexton II. Later production Sextons also featured the heavy duty vertical volute suspension with the return roller mounted on an arm to the rear of the bogie assembly. The 17-tooth drive sprocket and Canadian dry pin tracks were fitted to all but the earliest Sextons. (PAM)

The driver was seated down in the hull to the right of the transmission housing. The instrument panel was located directly in front of the driver. Twin control levers were used to steer the vehicle. (Author)

Ammunition was stored in bins under the fighting compartment floor. These bins could hold 112 projectiles and cartridges. (Author)

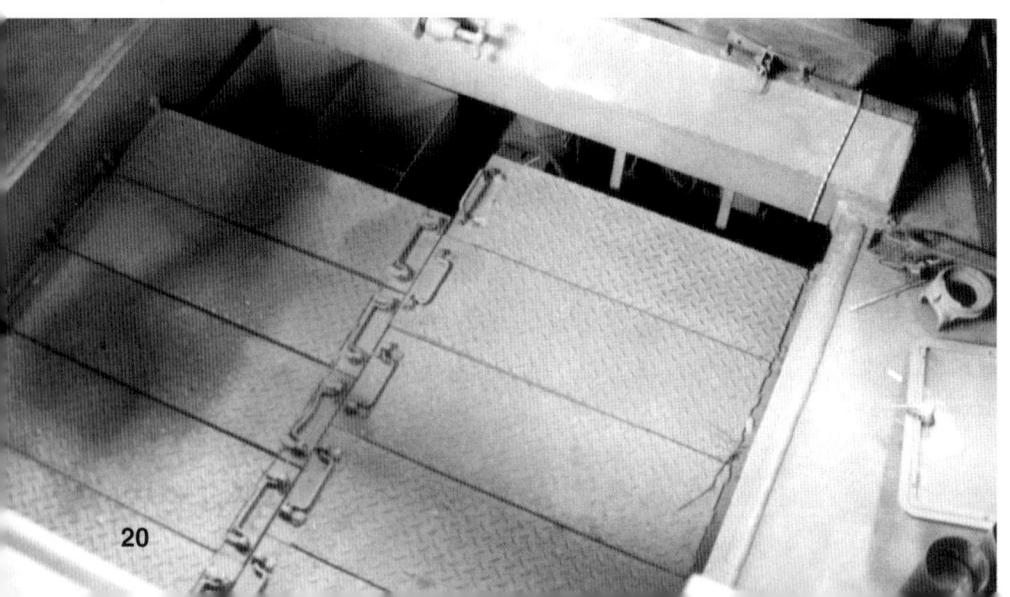

M7/Sexton

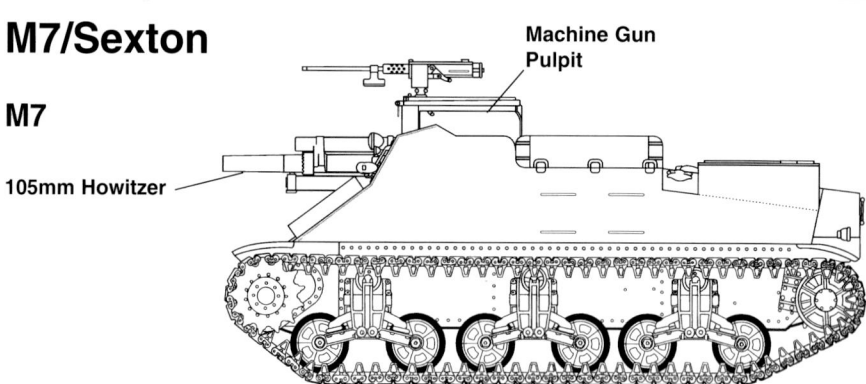

M7

105mm Howitzer — Machine Gun Pulpit

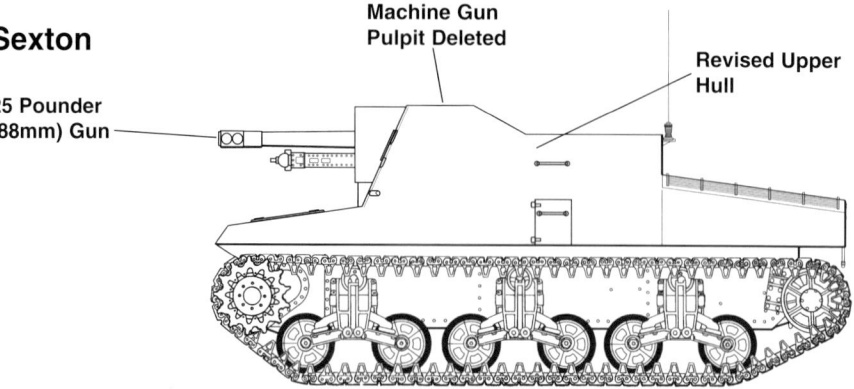

Sexton

25 Pounder (88mm) Gun — Machine Gun Pulpit Deleted — Revised Upper Hull

Additional ammunition storage bins were located in the right rear of the Sexton's fighting compartment. A 25 pounder projectile and three charges are resting on the floor at left. (Author)

A No. 19 Mk II or III radio set was located in the left rear corner of the fighting compartment. A wire screen provided protection against equipment accidentally damaging the set. Sten sub-machine guns were placed in storage clips on the port side of the fighting compartment forward of the radio. (Author)

The auxiliary generator box was located on the right side of the rear deck. The bracket in front of this box was for a telephone cable reel. Wire loops were welded to the engine deck to tie down stowage. The Sexton II also had rails along the sides and the rear of the engine deck to further assist with equipment stowage. (Author)

A South African Sexton fires on German positions in northern Italy during March of 1945. The 25 pounder gun could elevate to +40°, 5° more than the M7's 105mm howitzer. The crew improvised a wire frame for holding a tarpaulin over the fighting compartment to provide protection against the sun and the elements. (PAM)

M8 Howitzer Motor Carriage

In late 1941 the Armored Forces Board requested the development of a vehicle mounting a 75mm howitzer to provide fire support for mechanized units. The Field Artillery had performed some work in this area during 1939 and 1940 using the chassis of the M1 Combat Car. This work resulted in a vehicle designated the T3 Howitzer Motor Carriage (HMC). The T3 featured a 75mm M1A1 pack howitzer mounted in the starboard front hull. Additionally, a .30 caliber turret mounted machine gun was placed on top of the hull to the left of the vehicle centerline. This configuration was similar to that of the M3 Lee medium tank. The T3 HMC was judged unsuitable for the Field Artillery's needs as a support weapon primarily due to insufficient space for the three man crew. The Armored Forces Board briefly considered the T3 for use as an assault gun, however, no additional vehicles were produced.

Data from the T3 evaluation was used to develop the T18 HMC which employed the M3 Stuart light tank chassis. Two soft steel prototypes were manufactured by the Firestone Tire and Rubber Company for testing purposes. The T18 was also similar in layout to the M3 medium tank, although it lacked a turret on the upper hull. The T18 was equipped with a 75mm M1A1 howitzer mounted in the starboard front corner using a modified gun mount from the M3 Lee medium tank. The T18 HMC was completed in May of 1942, however, it was not accepted for service due to the development of superior vehicles.

While development of the fully tracked T18 was taking place, the T30 HMC was rushed into service as an interim measure. The T30 was based on the existing M3 Half-Track and mounted a 75mm M1A1 howitzer. Five hundred T30s were completed during 1942. The T30s provided good service, however, a better vehicle was needed. A new HMC would require improved armor protection for the crew and engine — both of which were lacking in the T18 and T30.

The new M5 Stuart light tank was used for two additional prototypes studied by the Armored Force in early 1942. The first prototype, designated the T41 HMC, mounted a 75mm howitzer in a modified open topped M5 chassis. The vehicle was studied during the spring of 1942, but it did not proceed beyond the mockup stage. The T41 was soon canceled in favor of the T47 HMC.

The T47 had initially been based on the M3 Stuart light tank chassis, however, this was changed to the later M5 chassis. The M5 used the suspension components of the M3, but featured a redesigned hull for improved ballistic protection and crew space. Additionally, the M5 was powered by twin gasoline engines in lieu of the single gasoline or diesel powerplant of the M3. The T47 featured a 75mm M2 howitzer in a fully rotating turret versus the limited traverse mount used in the earlier T3, T18, T30, and T41 HMCs. The M2 howitzer used the tube and breech assembly of the earlier M1A1 howitzer which was fitted into a new gun mount. This mount used components from the M34 mount employed on the M4 Sherman medium tank.

The T47 was ready for testing in early 1942. The T47 was redesignated the 75mm M8 Howitzer Motor Carriage and was sent to Aberdeen Proving Ground, Maryland, Fort Knox, Kentucky, and the Desert Training Center in California for trials during the spring of 1942. These trials resulted in only inconsequential modifications. Production began at the Cadillac Division of General Motors during September of 1942. By the time production ended in January of 1944, 1778 M8 HMCs had been produced.

Apart from the turret, the M8 HMC was similar to the M5 light tank. The M8's turret ring was eight inches (20.3 cm) wider than the turret ring of the M5, resulting in more interior room for the crew and weapon. The crew had greatly improved visibility in the open topped turret, although this came at the cost of overhead protection. The upper hull hatches on the M5 were eliminated on the M8 HMC due to the open turret. These hatches were replaced by large vision hatches in front of the two drivers. The other two crew members rode in the turret — the gunner on the port side and the loader on the starboard side. A .50 caliber M2 machine gun was fitted into a ring mount at the starboard rear of the turret for close-in and anti-aircraft defense. The .30 caliber hull machine gun fitted to the M5 was deleted on the M8.

The 75mm howitzer had a 360° arc of fire and an elevation range from -20° to +40°. The how-

The T3 Howitzer Motor Carriage appeared in 1939 and mounted a 75mm M1A1 pack howitzer on the chassis of the M1 Combat Car. When the doors around the howitzer were closed the weapon could only be elevated from -10° to +20°. When the doors were opened the howitzer could also be traversed 10° to port and 15° to starboard. (PAM)

The T18 used the M3 Stuart light tank chassis and mounted the 75mm M1A1 pack howitzer on the starboard side of the closed top fighting compartment. The weapon used a modified version of the 75mm M1 gun mount used in the M3 Lee medium tank. The T18 was canceled in May of 1942 in favor of the T47. (PAM)

The M5 Stuart light tank employed the suspension components of the M3 Stuart. The newer M5 had a redesigned hull for improved ballistic protection and crew space. The M5 was powered by twin gasoline engines.

The M8 featured two large vision hatches on the glacis plate. These hatches replaced the two overhead driver's hatches found on the M5 light tank. The M8 turret ring, eight inches wider than that of the M5, provided increased operating space within the turret. A flash deflector surrounded the barrel of the 75mm howitzer. (PAM)

itzer could fire high explosive (HE), high explosive anti-tank (HEAT), and chemical smoke rounds out to a maximum range of 9500 yards (8687 meters). A total of 46 rounds of ammunition were carried in the hull, although the M8 could also pull a small two-wheeled ammunition trailer to supplement the internal load.

Power for the M8 HMC was supplied by two liquid cooled Cadillac Series 42 engines, providing a total of 295 horsepower. The engines were equipped with Hydra-Matic automatic transmissions. These engines provided the M8 with a maximum speed of 36 miles (58 km) an hour. Cruising range was 100 miles (161 km) on 89 gallons (337 liters) of gasoline. The M8 weighed 34,600 pounds (15,694 kilograms) fully loaded. Early production M8s were equipped with open spoked road wheels, however, these were replaced with solid steel road wheels on later production examples.

The M8 HMC began to arrive in the combat zone during late 1943 and saw service in both

The T47 used the chassis of the improved M5 Stuart light tank and featured a 75mm M2 howitzer mounted in a fully traversing open topped turret. The M5 chassis combined the vertical volute spring suspension of the M3 light tank with a redesigned hull for improved ballistic protection and crew space. The T47 was redesignated the M8 Howitzer Motor Carriage during the spring of 1942. (PAM)

Track grousers were fitted to the sides of the M8's turret where they provided a small degree of spaced armor protection. Grousers were clipped to the tracks to provide increased traction under poor ground conditions. (PAM)

The large, open topped turret had a ring mounted .50 caliber M2 machine gun at the starboard rear corner. The gunner's direct sight telescope is mounted to the right of the howitzer. Pioneer tools were stowed on the back of the hull. (PAM)

The 75mm M2 howitzer used the barrel and breech assembly of the earlier M1A1 pack howitzer. The new weapon used components from the T34 mount employed on the M4 Sherman medium tank. The tubes on either side of the howitzer are part of the recoil mechanism. The gun assembly and the inside of the turret were painted Olive Drab. (PAM)

Italy and Western Europe. The M8 equipped the Headquarters (HQ) Companies of medium tank battalions until the arrival of M4 Sherman medium tanks fitted with 105mm howitzers during the spring of 1944. Additional M8s were supplied to the Free French Army for fire support duties in their armored divisions.

The M8 HMC was removed from the US inventory after World War Two. Surplus vehicles were supplied to various allies under defense aid agreements. M8s were used extensively by the French in Indochina during the late 1940s and early 1950s. The fledgling South Vietnamese Army received a number of these vehicles from the departing French after the Indochina conflict ended and continued to operate M8 HMCs until the early 1960s.

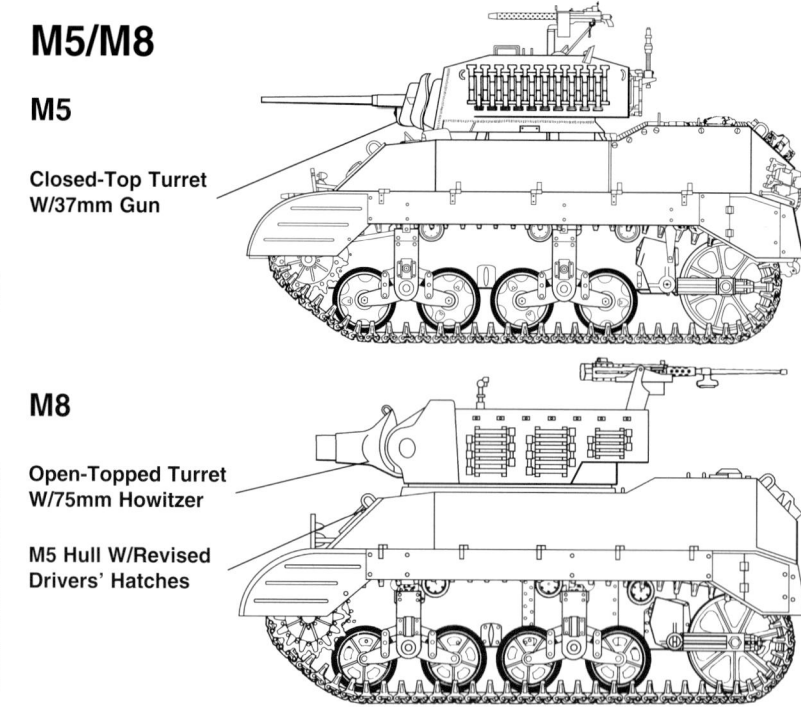

M5/M8

M5

Closed-Top Turret W/37mm Gun

M8

Open-Topped Turret W/75mm Howitzer

M5 Hull W/Revised Drivers' Hatches

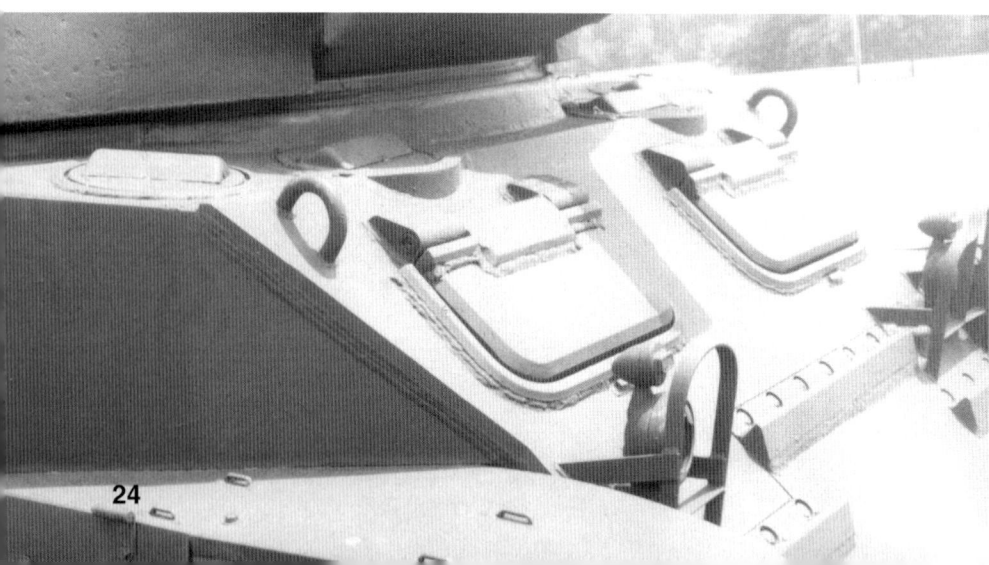

The vision hatches on the hull front could be opened up to 30° — parallel to the top of the hull. Each driver had two periscopes on the hull roof for forward and side vision when the hatch was closed. Removable windshields provided protection against the elements when the hatches were open. (Author)

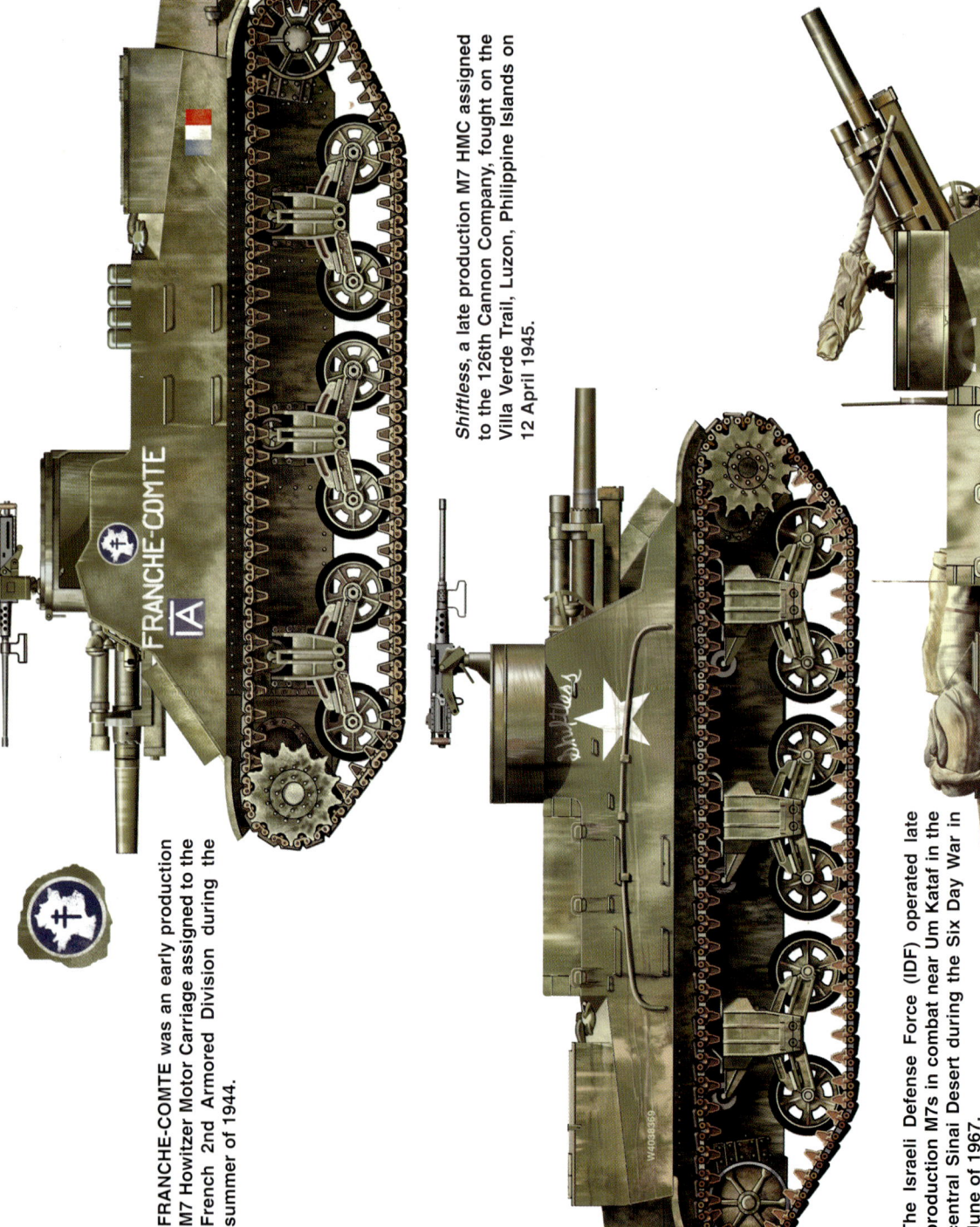

FRANCHE-COMTE was an early production M7 Howitzer Motor Carriage assigned to the French 2nd Armored Division during the summer of 1944.

Shiftless, a late production M7 HMC assigned to the 126th Cannon Company, fought on the Villa Verde Trail, Luzon, Philippine Islands on 12 April 1945.

The Israeli Defense Force (IDF) operated late production M7s in combat near Um Kataf in the central Sinai Desert during the Six Day War in June of 1967.

This M8 Howitzer Motor Carriage, towing an M8 Ammunition Trailer, served with an unidentified unit in Italy during November of 1943. The M8 trailer carried an additional 96 rounds of 75mm howitzer ammunition.

Winter-camouflaged M8s were assigned to 6th Reconnaissance Squadron, 6th Cavalry Group, during the fighting at Bastogne, Belgium in December of 1944.

The French 1er Regimént de Chasseurs a' Cheval operated M8s between Son Tay and Trung Ha, Tonkin, Indochina during 1951.

June Gil/Avant le Char de Mort (Forward, Tank of Death), an M12 Gun Motor Carriage serving with the 987th Field Artillery Battalion, supported the US attack on St Lo, France on 16 July 1944.

COURAGEOUS CONFEDERATE was an M40 GMC assigned to the 937th Field Artillery Battalion in central Korea during the winter of 1952-53.

(Above) The top and the bottom of the breech was surrounded by tubular breech guards. The direct vision sight and the elevation and traverse control wheels were located to the right of the breech guard. The firing button for the 75mm howitzer was installed just ahead of the breech guard assembly. The box to the left of the breech mechanism held .45 caliber ammunition for the crew's sub-machine guns. (PAM)

(Above Right) The M8 turret rotated over the open crew compartment. The gunner and loader seats were attached to the turret ring. Forty-six rounds of 75mm ammunition were stowed in the rear of the crew compartment, side sponsons, and forward hull. Two .30 caliber M1 carbines are stored on racks next to the gunner's seat. (PAM)

(Right) Six ready rounds of 75mm ammunition were vertically stowed in the forward hull between the drivers. Only the driver had an instrument panel, however, both drivers' positions were equipped with a set of controls. Periscopes were mounted above each driver's position. A fire extinguisher was installed in the front of the assistant driver's position. The drivers' compartment was painted white to improve the interior lighting. (PAM)

(Above) A 1st Armored Regiment M8 bypasses a destroyed bridge near Veiano, Italy on 8 June 1944. Sandbags and track links were added to the glacis plate for additional protection. This vehicle is towing the standard M8 ammunition trailer which held an additional 96 rounds of 75mm howitzer ammunition — almost tripling the ammunition on the HMC. The M8 HMC had replaced the half-track T30 HMC in the fire support role by October of 1943. (US Army/NA)

(Below) An M8 from the 33rd Armored Regiment, 3rd Armored Division sits in the town of Marigny, France on 26 July 1944. The prominent star and circle on the turret roof was an air identification aid to avoid attacks by Allied pilots. This M8 has additional gear stowed on the fenders. A steel towing cable has been attached to the port towing clevis on the differential housing. The cable is running down the left side of the M8. (US Army/NA)

(Above) The reconstituted French Army under General Charles de Gaulle received large quantities of American equipment — including M8s. The crew of this M8 from the French 2nd Armored Division is welcomed by the citizens of Maintenon, France on 23 August 1944. The division was passing through Maintenon on its way to liberating Paris two days later. The glacis plate is marked with a Cross of Lorraine, the emblem of the Free French forces. (US Army/NA)

(Below) M8s from the 113th Cavalry Squadron, 2nd Armored Division prepare to fire on the retreating Germans at Heure-Le-Remam, Belgium on 9 September 1944. The near M8 has been fitted with the Culin hedgerow cutter which was unusual for this type of vehicle. The cutter was designed to help punch a hole in the thick hedgerow walls which crisscrossed the Normandy battlefield. This vehicle is also equipped with the later solid steel road wheels. (US Army/NA)

(Above) Two M8s, LITTLE WEST POINT and MIGHTY, from the 15th Cavalry Group, 94th Infantry Division, prepare to fire in support of a cavalry reconnaissance patrol near Blain, France on 23 December 1944. MIGHTY is equipped with a Culin hedgerow cutter and has spare track links attached to the glacis plate for additional armor protection. (US Army/NA)

Rubber Block Track with Grouser

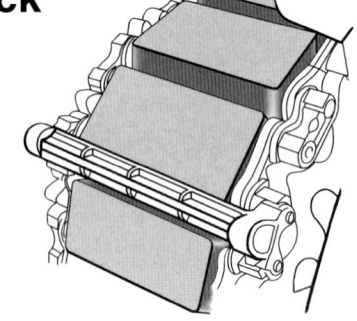

(Left) An M8 of Troop E, 106th Cavalry Group fires on enemy positions in Geislautern, Germany on 8 February 1945. M8 crews usually threw spent 75mm howitzer shells out of the vehicle immediately after firing. Grousers have been added to the T16 rubber block tracks to improve traction. One grouser is visible just in front of the drive sprocket. This M8 is equipped with late production pressed steel road wheels, however, it retains the early production drive sprocket and idler wheel. (US Army/NA)

(Above) The M8 was well-liked for its ability to fire at targets emplaced on hilly terrain. Three M8s of the 758th Light Tank Battalion fire on German positions near Seravezza, Italy on 8 April 1945. This fire was supporting the famous Nisei (first generation Japanese-American) 442nd Infantry Regiment — the most highly decorated unit in the US Army during World War Two. The near M8 is fitted with the early style spoked wheels rather than the more common pressed steel type. (US Army/NA)

(Right) The M8 saw extensive service with French forces in Indochina following World War Two. The crew of this vehicle cover a road in the Tonkin region of northern Vietnam during 1952. The insignia on the turret is from the *1er Regiment de Chasseurs*. A towing cable has been looped around the turret and front hull of this M8. (ECPA via Balin)

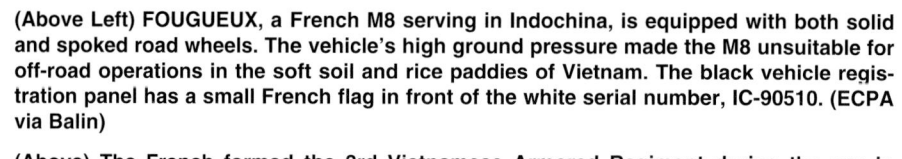

(Above Left) FOUGUEUX, a French M8 serving in Indochina, is equipped with both solid and spoked road wheels. The vehicle's high ground pressure made the M8 unsuitable for off-road operations in the soft soil and rice paddies of Vietnam. The black vehicle registration panel has a small French flag in front of the white serial number, IC-90510. (ECPA via Balin)

(Above) The French formed the 3rd Vietnamese Armored Regiment during the war in Indochina. The 3rd VAR remained in service with the new South Vietnamese Army following the 1954 Geneva peace accords. These Vietnamese troops are training with an M8 from the Regiment's 3rd Squadron. (ECPA via Balin)

(Left) The end of the line. The author discovered this abandoned South Vietnamese M8 in a compound near Gia Dinh during his tour in Vietnam in early 1972. These were the last M8s to see combat — almost thirty years after they first saw action in Italy in late 1943. (Author)

M12 Gun Motor Carriage

The US Army Ordnance Department initially performed some development work on a self-propelled 155mm gun following World War One, however, the Field Artillery Branch (FAB) showed little enthusiasm for such a weapon. The FAB believed that either horses or tractors were sufficient to meet the Field Artillery's towing needs. This attitude slowly changed with the advent of World War Two in Europe. Funding became available to re-equip and modernize the small American army in anticipation of being drawn into the conflict. Development work began during June of 1941 on a 155mm self-propelled gun mount based on the new M3 Lee medium tank chassis. The Ordnance Department directed this self-propelled gun project while working in conjunction with the Artillery Branch. A pilot vehicle, designated T6, was ordered constructed at Rock Island Arsenal in Illinois.

The T6 used the 155mm M1918M1 gun with its recoil mechanism and carriage mounted in the hull of a modified M3 chassis. The gun was mounted in the rear of the hull. This required the relocation of the engine from the vehicle rear to the hull center. The repositioned engine resulted in a shorter transmission shaft. A hydraulically operated recoil spade was mounted at the rear of the T6 to stabilize the vehicle during firing. The driver and assistant driver/co-driver sat in the hull front on either side of the transmission. Both were equipped with roof and side access hatches. Each driver was also provided with a small hinged vision flap fitted with a protectoscope. The vision flap was nearly identical to that found on the M3 medium tank. The T6 used the early vertical volute suspension with the return roller mounted above the center of the bogie assembly.

The T6 pilot arrived at Aberdeen Proving Ground, Maryland on 12 February 1942. Tests at Aberdeen demonstrated that the T6 was a stable firing platform, however, the recoil spade experienced some hydraulic problems. The spade also protruded too much when in the retracted position which limited cross-country mobility. After the recoil spade was modified, the T6 was shipped to Fort Bragg, North Carolina for additional tests. The T6 quickly demonstrated its superior mobility and emplacement time.

The US Army was satisfied with the test results and ordered the T6 into production. The vehicle was redesignated the 155mm M12 Gun Motor Carriage (GMC). Fifty M12s were initially ordered from Pressed Steel Car Company during the summer of 1942. This order was later increased to 100 vehicles in August of that year and the first production vehicles began coming off the assembly line in September of 1942. The final M12 GMC was delivered during March of 1943.

The first production M12 GMC was subjected to additional testing at Erie Proving Ground, Pennsylvania and at Fort Bragg. Data from these tests resulted in engine modifications to eliminate a vapor lock problem. Additional modifications were made to the engine flame arrestor and exhaust pipes.

Throughout 1943 the Artillery Branch displayed little interest in employing the M12 GMC. When M12s came into service they were sent to training units or put into storage. By 1943, however, US Army plans for the invasion of Europe were coming to fruition. Test and training reports led the Army to upgrade and modernize the M12s in December of 1943. The original M3 vertical volute suspension was replaced by the heavier duty M4 suspension with rear mounted return roller. A small shield was mounted on the gun mount and the co-driver's side hatch was deleted. The modifications were finalized during February of 1944 and the Baldwin Locomotive Works began remanufacturing 74 M12s to the new standards. M12 remanufacturing work was completed by May of 1944.

The 155mm gun used on the M12 GMC was based on the French 155mm gun used during World War One. The M12 employed either the M1917, M1917A1, or M1918M1, depending on their availability. The M1917 gun was made in France and featured a French breech ring,

The T6 pilot mounted a 155mm M1918M1 gun on an M3 Lee medium tank chassis. During service tests the T6 demonstrated the superiority of self-propelled artillery over towed artillery for supporting armored units. The 155mm gun has been placed in a travel lock to reduce strain on the elevation mechanism during vehicle movement. (PAM)

The recoil spade protruding at the rear of the T6 inhibited cross country mobility and required design changes as a result of testing. Both the driver and assistant driver had side access doors. The T6 featured the early vertical volute suspension with open spoked wheels and return rollers mounted above the center of the bogie assembly. (PAM)

(Above Left) Both the T6 and the production M12 Gun Motor Carriage used the three piece transmission housing of the early production M3 chassis. Both drivers were protected by their enclosed positions, however, there was little protection for the gun or the remaining crewmen. The T6 also featured the T41 rubber block track. (PAM)

(Above) During December of 1943, the US Army decided to modify the M12 in anticipation of the upcoming invasion of Europe. The heavy duty vertical volute suspension with rear mounted return rollers used on the M4 replaced the earlier M3 suspension components. The assistant driver's hatch on the starboard side of the hull was also deleted. The modernized M12 retained the older three piece transmission cover. (PAM)

while the M1918M1 was completely American made. The M1917A1 was the French made M1917 fitted with the American made breech used on the M1918M1. The 155mm gun on the M12 had an elevation of -3° to +35° and could traverse 28° — 14° to either side. Ten rounds of ammunition and propellant charges were carried in the rear fighting compartment. The gun could fire high explosive (HE), armor piercing, white phosphorous (WP), and chemical smoke rounds. Maximum range of the gun was 15,200 yards (13,382 meters) with a normal propellant charge or 18,700 yards (17,017 meters) with a supercharge.

The M12 GMC was operated by a crew of six men. Power was provided by a 353 horsepower, air cooled Continental R975 radial engine. This engine provided a maximum speed of

(Left) The M12 featured a small gun shield on the port side of the 155mm gun, however, the gun crew remained relatively exposed. This M12 has been fitted with the T51 non-reversible rubber block track which was thicker than the earlier T41 reversible track. The T51 track provided excellent traction with little damage to hard roads, however, this track could slip when climbing over soft terrain. (PAM)

(Above) During initial testing the recoil spade protruded from the hull rear, a design feature which limited the vehicle's cross country movement. The spade was redesigned to correct this problem. The hydraulically actuated recoil spade allowed production M12s to be quickly emplaced with little preparation. (PAM)

(Above Right) Four of the M12's six crewmen sat in the open rear fighting compartment. Two crewmen were seated to the left of the gun, while the remaining two crewmen sat on the retracted recoil spade. The M12 lacked a machine gun for self protection; close-in defense was provided by the crew's personal weapons. (PAM)

24 miles (38.6 km) per hour. A fuel capacity of 200 gallons (756 liters) of gasoline provided a maximum range of 140 miles (225.4 km).

Due to the small ammunition load of the T6/M12 GMC, the US Army decided to construct a companion ammunition carrier and service vehicle. This vehicle, initially designated the T14, was nearly identical to the T6/M12 except for the removal of the 155mm gun and recoil spade. Removing the gun resulted in stowage space for forty rounds of ammunition and propellant charges. A ring mounted .50 caliber M2 machine gun was added to the center of the rear fighting compartment for self defense. The T14 was redesignated the M30 Cargo Carrier and 100 vehicles were manufactured by the Pressed Steel Car Company between October of 1942 and March of 1943. The Baldwin Locomotive Works modernized 74 M30s to the same standards as the M12 GMC. The M30 Cargo Carrier was attached to the M12 units then in Europe.

(Right) When the M12 entered service in September of 1942 it was used for training or put into storage. This M12 crew swabs out the 155mm gun barrel prior to loading during training at Camp Bowie, Texas on 30 September 1943. (US Army/NA)

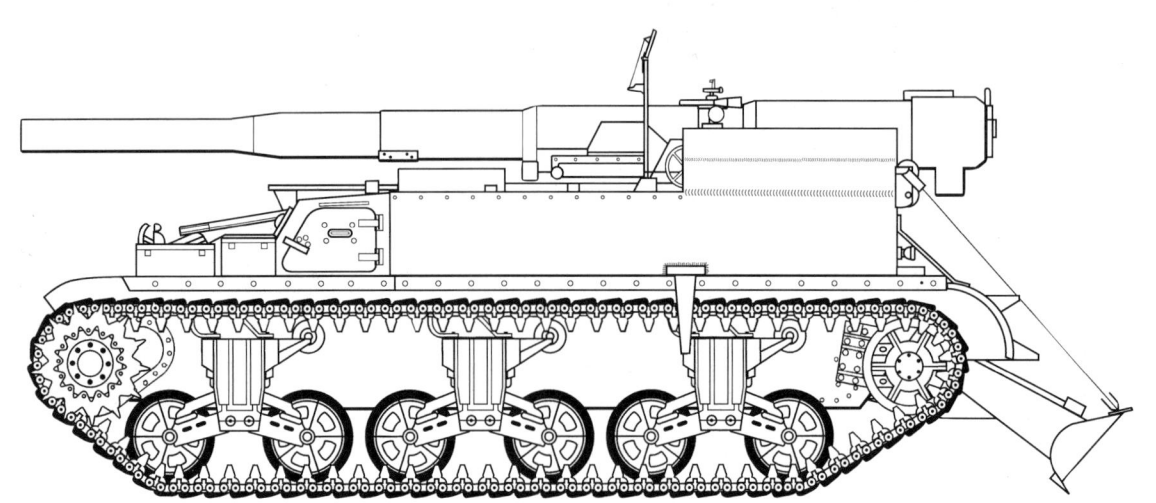
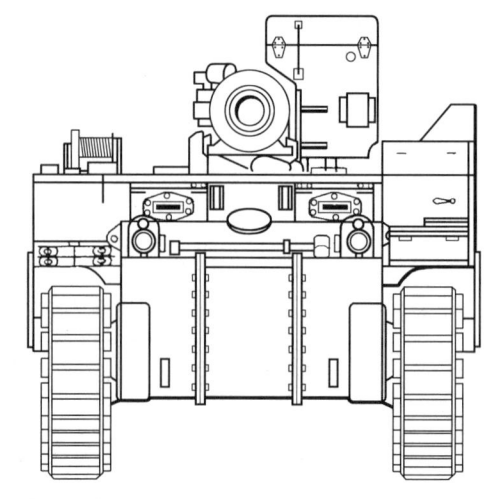

M12 GMC Specifications

Length................22 Feet, 1 Inch
Width..................8 Feet, 9 Inches
Height.................9 Feet, 5.5 Inches
Weight...............59,000 Pounds
Armament.........1 x 155mm M1917, M1917A1, or M1918M1 gun
Speed.................24 MPH
Crew...................6

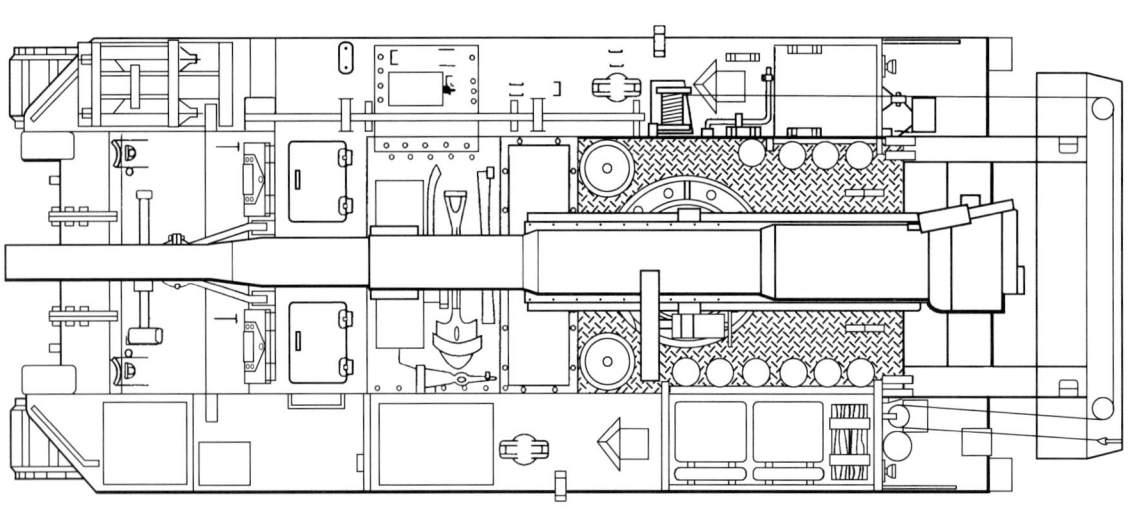
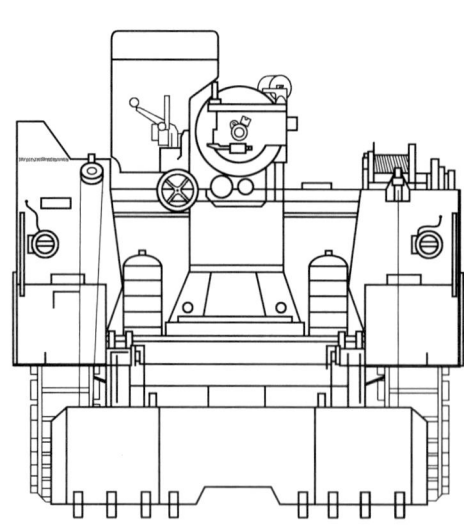

(Above) CORREGIDOR, an M12 assigned to the 987th Artillery Battalion, fires against German positions around St Lo on France's Contentin peninsula on 16 July 1944. The 987th AB was the first unit to take the M12 into action during OPERATION COBRA — the breakout from the Normandy beachhead. A camouflage net covers the rear of the M12. (US Army/NA)

Combat Service

The M12 Gun Motor Carriage had been assigned to six artillery battalions by the time of the invasion of France on 6 June 1944. These battalions were deployed to the beachhead area and took part in the breakout and drive across France that summer. Due to the speed of the Allied advance the M12 was often the only heavy artillery available for fire support missions. The M12 proved to be a deadly and well-liked weapon by American forces.

When the fighting neared the German border, the Allies ran into the vaunted Siegfried Line — the German counterpart to the French Maginot Line. Although the Siegfried Line was not as strong as it was purported to be, it was nevertheless a formidable obstacle. The bunkers of the Siegfried Line proved to be tough nuts to crack when manned by motivated German troops. Under these circumstances the M12's 155mm guns were used in the direct fire role, hitting bunkers and pillboxes at ranges of up to 2000 yards (1820 meters). The

(Right) The crew of *Adolph's Assassin*, an M12 assigned to the 991st Field Artillery Battalion, fires their gun against German positions near Kornelmunster, Germany on 4 November 1944. The running gear, tracks, and recoil spade are covered in mud from off-road travel. (US Army/NA)

(Above) A crew from Battery C, 575th Field Artillery Battalion, demonstrates their M12 for visiting officers under snowy skies in the Morteau sector of France on 15 November 1944. To help stabilize the gun, chocks have been placed under the front of the tracks to aid the dug-in recoil spade. This vehicle is fitted with the late production drive sprocket and solid idler wheel, but retains the early open spoked road wheels. The segmented ring around the star was due to the stencil pattern often not being filled in to complete the circle. (US Army/NA)

155mm high explosive rounds were able to penetrate up to seven feet (2.1 meters) of reinforced concrete when equipped with special concrete piercing fuses. The M12 proved to be an ideal 'bunker busting' weapon and was instrumental in breaking through the German line of fortifications.

M12 GMCs were in constant use throughout World War Two, however, when hostilities ended in Europe on 8 May 1945 the vehicles were declared obsolete. The M12s were scrapped, along with their companion M30 Cargo Carriers. The limited number of M12s prohibited them from being supplied to US allies during World War Two. Additionally, M12s did not see action in the Pacific Theater.

(Left) The crew of BUCCANEER has added sandbags to the front of their vehicle for additional protection. Sandbags were not normally seen on self-propelled artillery pieces, being more associated with tanks and tank destroyers. This M12 is equipped with T48 rubber chevron tracks — the most common track type used on the M4 Sherman series of vehicles. (US Army/NA)

ALBERTA IV and other M12s of the 11th Armored Division fire on German positions near Budesheim, Germany on 10 March 1945. This division was part of the US Third Army, commanded by General George S. Patton, Jr. Two jerry cans were placed on the left fender of ALBERTA IV. Tents for sheltering the M12 crews were pitched behind their vehicles. (US Army/NA)

An M12 prepares to fire on a German pillbox across the river from Echternach, Luxembourg on 8 February 1945. The M12 proved to be an ideal weapon to knock out the heavy pillboxes of the Siegfried Line on the French-German border. The 155mm rounds, when equipped with special concrete piercing fuses, could penetrate up to seven feet of reinforced concrete when used in the direct fire role. The lack of overhead armor, however, made the crews and gun vulnerable to counter artillery, mortar, and small arms fire. (US Army/NA)

Pill-Box ANNIE, an M12 of the 989th Field Artillery Battalion, 45th Infantry Division, fires on a Siegfried Line pillbox near Seelbach, Germany on 18 March 1945. The 155mm gun is in full recoil and has kicked up a large amount of dust in front of the vehicle. The recoil has caused the front of the M12 to rise slightly off the ground. (US Army/NA)

A 9th Armored Division M30 Cargo Carrier moves down a road toward Westerengel, Germany on 12 April 1945. The M30 is towing a barely visible M8 Armored Trailer. M30s were attached to M12 units where they served as ammunition and service vehicles. (US Army/NA)

The M30 cargo carrier was developed from the M12 Gun Motor Carriage. The 155mm gun and recoil spade were deleted to allow the M30 to carry up to 40 rounds of ammunition and propellant charges — four times the ammunition capability of the M12. The M30 was armed with a .50 caliber M2 machine gun for close-in defense. M30s were assigned to M12 units in Europe. (US Army)

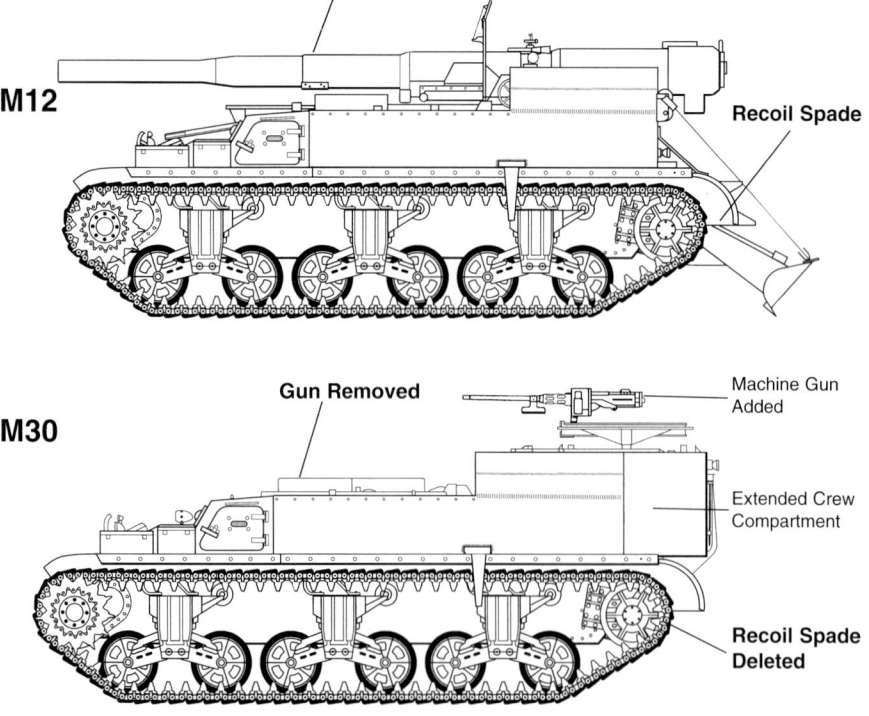

M12/M30

M40 GMC/M43 HMC

Due to the success of the M12 Gun Motor Carriage (GMC) design, the Artillery Branch requested additional vehicles. Unfortunately there were no additional M1918M1 guns available for the conversion. The newer 155mm M1 gun was considered, however, studies indicated that the heavier and more powerful M1 weapon would overstress the M12 chassis. The US Army then decided to mount the M1 gun in the more robust M4 Sherman medium tank chassis. This self-propelled gun would use as many up-to-date M4 components as possible. The hull of the new vehicle was wider than that of the M12 GMC — 124 inches (315 cm) versus 105.3 inches (267.5 cm). The newer horizontal volute spring suspension (HVSS), introduced on the M4A3E8 medium tank, was used in lieu of the older vertical volute suspension system. The general outline of the new self-propelled mount resembled the M12 with the driver and co-driver in the front of the hull, the engine placed in the center, and the gun mount located in the rear. Five pilot vehicles, designated the 155mm T83 GMC, were ordered in March of 1944 from the Pressed Steel Car Company. The first T83 was completed by the end of July.

Test firing of the T83 was conducted at Aberdeen Proving Ground where 200 rounds were expended during the initial series of tests. The results were good under all conditions. The trials also showed that the gun could be fired with or without the recoil spade emplaced. After these tests, the 8-inch (203mm) M1 howitzer replaced the 155mm M1 gun for further evaluation. These howitzer tests also produced good results. The last two T83s were then rearmed with this weapon and redesignated the 8-inch T89 Howitzer Motor Carriage (HMC). Further testing of the T83 resulted in minor changes to the interior and stowage arrangements. The US Army was satisfied with the design and ordered the T83 into production in January of 1945. In May of that year the T83 was redesignated the 155mm M40 Gun Motor Carriage. Four hundred eighteen M40s had been manufactured by the Pressed Steel Car Company when production ceased at the end of 1945. Twenty four M40 GMCs were later converted into T89s.

The two T89 pilot vehicles were subjected to further tests during January of 1945. The T89 was designed to be a universal chassis able to mount either the 155mm M1 gun or the 8-inch M1 howitzer and their associated ammunition racks. An order for 576 T89s was originally given to Pressed Steel, but this order was reduced to 48 vehicles due to the end of World War Two. The production total included the 24 M40s converted to T89s. The T89 was redesignated the 8-inch M43 Howitzer Motor Carriage in November of 1945.

Both the M40 GMC and M43 HMC were designed around the basic M4A3 chassis with the horizontal volute suspension and one-piece nose casting. The driver and co-driver sat in the front of the hull. Each driver's position had his own cupola and hatch. A small floor mounted escape hatch was placed behind the assistant driver's position. A travel lock for the gun or howitzer was mounted on the sloped glacis plate.

The air-cooled Continental R975 nine cylinder engine was mounted directly behind the drivers in the middle of the hull. This 960 horsepower engine propelled both the M40 and M43 to a maximum speed of 24 miles (38.6 km) per hour. Range for both vehicles on 215 gallons (813.9 liters) of gasoline was 100 miles (161 km). Both the M40 and M43 were operated by an eight-man crew.

Total traverse for both the 155mm gun and 8-inch howitzer was 36° — 18° to either side of the vehicle's centerline. Both weapons had an elevation range of -5° to +45°. The M40 could carry 20 rounds of 155mm ammunition, while the M43 could carry 16 rounds of 8-inch ammunition. The 155mm gun could fire high explosive (HE), armor piercing, or white phosphorous (WP) rounds out to a maximum range of 25,715 yards (23,514 meters). The 8-inch howitzer could fire HE rounds out to a maximum range of 18,510 yards (16,926 meters). Each weapon fired separate loading ammunition consisting of projectiles and charges. The normal rate of fire for both weapons was one round per minute.

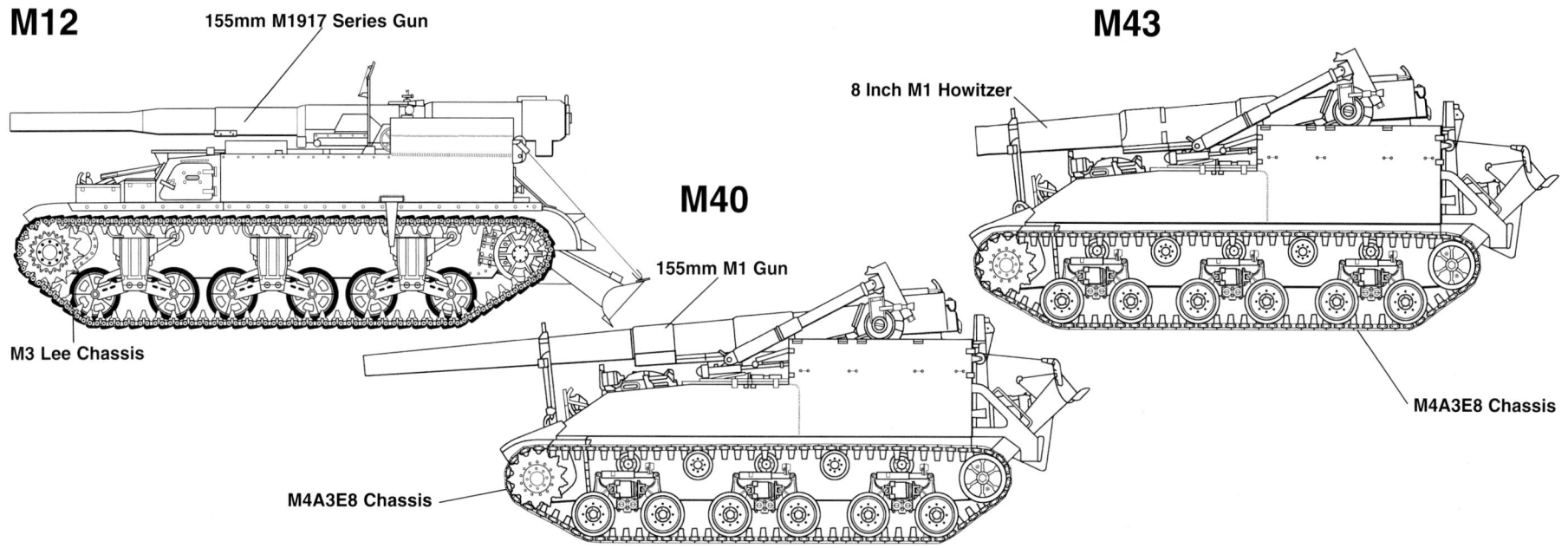

(Above Left) The 155mm M1 gun of the T83 Gun Motor Carriage was larger, heavier, and more powerful than the older M1918M1 gun employed on the M12. Consequently, the T83 used the stronger lower hull and horizontal volute suspension system of the M4A3E8 Sherman medium tank. (PAM)

(Above) The T83 featured the new horizontal volute spring suspension (HVSS) which provided a better ride and a lower ground pressure than that of the earlier vertical volute suspension. Unlike the older vertical volute suspension, the return rollers were mounted on the side of the hull. (Author)

Combat Service

One T83 and one T89 were assigned to the Zebra Mission during February of 1945. The Zebra Mission was responsible for evaluating new weapons under battlefield conditions in Europe. Both vehicles were assigned to the 991st Field Artillery Battalion, which was equipped with the M12 Gun Motor Carriage. The battalion immediately replaced the T89's 8-inch howitzer with a 155mm gun. This change allowed the T89 to fire the same ammunition used by the M12s in the 991st FA Battalion. The T83 and T89 were used in the attack on Cologne with satisfactory results. The 8-inch howitzer was then reinstalled on the T89 for further combat evaluation.

(Left) The T83 was redesignated the M40 GMC in May of 1945. The M40 used the one piece differential housing installed on later production M4s. Both drivers had their own hatches on top of the hull, however, the side hatches seen on the M12 were not present. A barrel travel lock and spare track links were carried on the glacis plate. (PAM)

(Above) The pilot T83 had the 155mm gun replaced by an 8-inch howitzer for testing. The ability to change the guns made the T83/T89 a versatile weapons system. The general configuration changed little between the pilot vehicles and the final production models. The T89 was redesignated the M43 HMC in November of 1945. (PAM)

(Above Right) The T89 Howitzer Motor Carriage was similar to the T83, but mounted an 8inch M1 howitzer in lieu of the 155mm M1 gun. This pilot vehicle is fitted with the T66 all steel cast track with a single pin connecting the track links. (PAM)

The head of the Zebra Mission, General Gladeon Barnes, returned to Aberdeen during the spring of 1945. He expressed concern about gun and crew protection for the new self-propelled mounts. When the M12s were used in the direct fire-support role, both the guns and crews were vulnerable to German artillery, mortar, and small arms fire. General Barnes directed Aberdeen to study the installation of secondary armament and overhead protection on both the T83 and T89.

A variety of weapons were mounted on both the T83 and T89 for evaluation. These included .30 caliber hull machine guns — identical to the weapon used on the M4 medium tank — and another in the rear fighting compartment. These weapons were not satisfactory due to traverse limitations. Recoilless rifles (57mm and 75mm) were also fitted into the fighting com-

The pilot vehicles lacked the spare track stowage on the glacis plate and had a different type of travel lock compared to the later production vehicles. The wider T66 cast steel tracks weighed 3000 pounds more than the tracks used on the M12, however, the T66 tracks provided the T83/T89 with a lower ground pressure, better traction, and a smoother ride. (PAM)

(Above) Two sets of spare tracks were carried on the glacis plate in front of the drivers' positions. The gun travel lock was mounted on the glacis plate between the spare track links. The travel lock kept the gun from moving while the vehicle was relocating. (Author)

(Above Left) The T83 was also equipped for deep wading with exhaust and intake extensions to prevent water from flooding the engine. The recoil spade support arms used I-beams instead of the solid arms used on the M12. (PAM)

partment for testing. The recoilless rifles proved unsatisfactory due to their powerful backblast being dangerous to the vehicle's crew. Aberdeen believed that .50 caliber machine guns would provide sufficient secondary armament for both the T83 and T89, however, the .50 caliber machine gun was not fitted to either vehicle.

A large armored cab was designed to provide more protection to the fighting compartment of both the T83 and T89. A full-scale mockup of an armored cab was installed over the fighting compartment of an M40 at Aberdeen. This arrangement proved unfeasible due to weight and configuration problems. Aberdeen recommended that the T83 and T89 be redesigned to provide more crew protection, however, this recommendation was not implemented.

Neither the M40 GMC or M43 HMC were available for large scale action before the end of

(Left) This is about as close as the US Army could get to having a battleship. The T83 could travel through 40 inches of water with the deep wading gear installed. It is believed that this capability was never employed since the M40 and M43 never participated in an amphibious assault. (PAM)

(Above) Each driver's cupola was equipped with six vision ports and a hatch periscope for indirect vision. This cupola was a standard feature on late production M4 turrets. (Author)

(Above Right) The rear of the M40 folded down to form a fighting platform. A sliding steel screen could be pulled out for additional working space. The ramp floor contained two seats for the crew members. (PAM)

World War Two. A limited number of both vehicles, however, were used to replace M12s which had reached the end of their useful life. When the Korean War broke out in June of 1950 the M40 and M43 were committed to the fighting. Both vehicles proved invaluable in the support role as the fighting stabilized along the 38th Parallel.

Following the Korean War, the M40 and M43 were gradually replaced by newer self-propelled artillery. Some of these newer pieces addressed the problem of overhead protection raised by the use of the unprotected M12 during World War Two. A small number of M40s were supplied to NATO allies under defense agreements. It is not known if any M43s were also supplied to these allies. By the early 1960s both the M40 and M43 had been replaced by more modern vehicles.

(Right) The recoil spade was raised and lowered using a winch and cables — a similar design to that of the earlier M12. Both the rear deck and fighting compartment floor had a raised non-skid tread pattern. The wire screen extension is retracted under the rear platform. (Author)

The M40 and M43 were equipped with either T66, T80, or T84 tracks. The T80 track is fitted to this M40 on display at the Patton Armor Museum at Fort Knox, Kentucky. This steel track had bonded rubber innerpads with narrow chevrons. (Author)

Two plates on the roof could be folded forward in front of the gun shields. The upper rear and side plates could be folded back to provide additional protection for the crew on the rear platform. (PAM)

The lack of overhead crew protection was a concern for the Zebra Mission commander, General Gladeon M. Barnes. Gen Barnes ordered a study made of ways to protect the T83's gun and crew. A full scale mockup of one design was installed over the fighting compartment of the M40 at Aberdeen Proving Ground, Maryland. (PAM)

The top and side plates are extended on this M40. Although this armored cab would have provided some crew protection, many parts of the gun were left unprotected. Aberdeen recommended redesigning the vehicle to provide adequate protection for both crew and weapon. This recommendation was not implemented. (PAM)

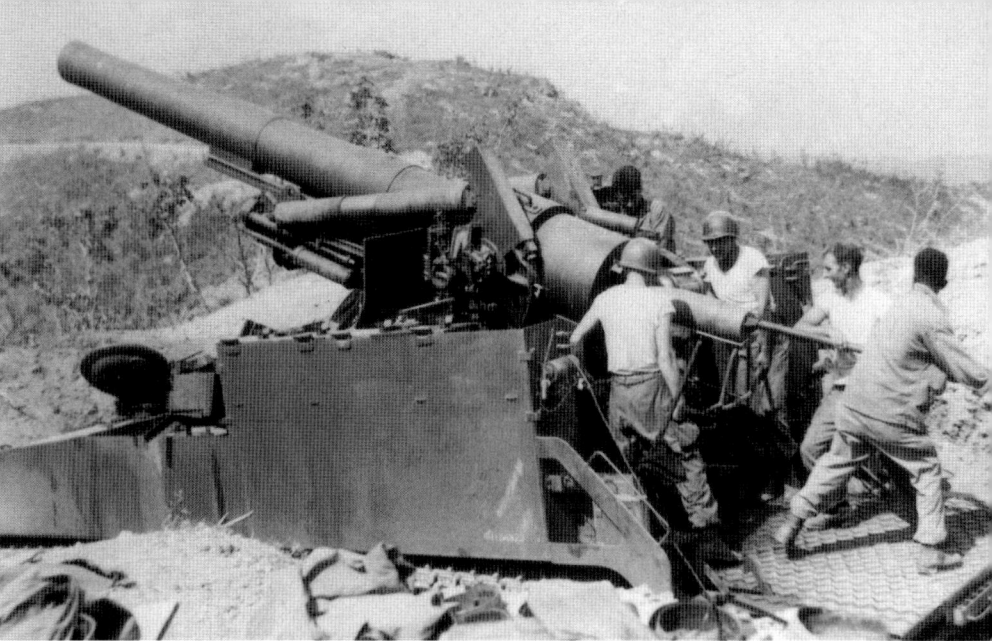

(Above) The M40 saw little service in World War Two, but was committed to the fighting in Korea. An M40 of Battery C, 204th Field Artillery Battalion, fires on Chinese positions north of Yonchon, Korea on 22 August 1951. The side of this M40 is emblazoned with a green and yellow dragon with red flames, heart, and trim. (US Army/NA)

(Below) Bunker-type artillery emplacements were constructed along the Korean 38th Parallel after the fighting stabilized. This M40 crew from 3rd Section, Battery B, 937th Field Artillery Battalion, cleans their gun on 5 January 1953. Several 155mm projectiles are vertically stored by the sandbag walls of the bunker. (US Army/NA)

(Above) Although less common than the M40, the M43 was also employed in Korea. This M43 crew rams a round into the breech during a fire mission during the summer of 1952. A loading cradle holding an 8-inch shell has been positioned just behind the breech. The shell, followed by a powder charge, is then pushed into the breech. (PAM)

(Above) CYD CHARISSE, an M40 assigned to Battery C, 937th Field Artillery Battalion, fires on Chinese positions in Korea on 5 January 1953. The crew is holding their ears for protection from the noise and concussion made by the 155mm gun. Two of the crewmen stand on the rear platform behind the gun. The wire screen extension to the rear platform is deployed. (US Army/NA)

(Below) One advantage of self-propelled artillery is the ability to quickly redeploy and set up for firing. BIG BRUISER/ARKANSAS LONG TOMS, an M40 from Battery B, 937th Field Artillery Battalion, digs in to help counter the Chinese breakthrough in the Republic of Korea Capital Division Sector on 14 July 1953. The 937th FA apparently based their vehicle names on battery letters. (US Army/NA)

(Above) The M40 was supplied in limited numbers to allied armies. This M40 was the first example received by the French under a NATO military aid program. The suspension and tracks are sagging due to the lack of weight on the vehicle. (US Army/NA)

M37/M41 Howitzer Motor Carriages

The development of the M24 Chaffee light tank during World War Two provided a new chassis for use as a self-propelled artillery mount.* Two pilot vehicles using the M24 chassis were developed for service before the end of the war.

The first pilot vehicle was designated the T76 and was similar in configuration to the M7 Howitzer Motor Carriage (HMC). A 105mm M4 howitzer was mounted in the front of the fighting compartment slightly to the right of the hull centerline. A ring mounted .50 caliber M2 machine gun for anti-aircraft and close-in defense was placed in a pulpit at the starboard front corner of the fighting compartment. The T76 was tested at Aberdeen Proving Ground, Maryland and Fort Knox, Kentucky beginning in July of 1944. Modifications resulting from the trials included increasing the 105mm ammunition capacity from 68 to 126 rounds. Additionally, the machine gun ring mount diameter was reduced from 42 to 36 inches (106.7 to 91.4 cm). The T76 was redesignated the M37 Howitzer Motor Carriage in January of 1945.

The 105mm M4 howitzer had an elevation from -10.5° to +42.8°, with a total traverse of 51.7° — 25.4° to port and 26.3° to starboard. The howitzer could fire high explosive (HE), high explosive anti-tank (HEAT), white phosphorous (WP), and chemical smoke rounds. Maximum range of the weapon was 12,205 yards (11,160 meters). Power for the M37 was supplied by two liquid cooled Cadillac Series 44T4 V-8 engines, which generated a combined 296 horsepower. The M37 was capable of a maximum speed of 35 miles (56.4 km) per hour. Maximum range on 110 gallons (416.4 liters) of gasoline was 100 miles (161 km). The M37 HMC carried a crew of seven men. Both American Car and Foundry and Cadillac produced a total of 316 M37s through October of 1945.

The second pilot vehicle was designated the T64E1. It was similar in layout to the M40 HMC and mounted a 155mm M1 howitzer in the rear of the hull. The T64E1 pilot vehicle was shipped to Aberdeen Proving Ground during December of 1944 for testing. Further trials were conducted at Fort Bragg, North Carolina beginning in January of 1945. Internal crew and equipment stowage was revised following these tests. Additionally, 155mm ammunition storage was increased from 18 to 22 rounds. The T64E1 was redesignated the M41 Howitzer Motor Carriage in June of 1945.

The 155mm M1 howitzer could elevate from -5° to +45°. Total traverse of the weapon was 37.5° — 20.5° to port and 17° to starboard. The 155mm howitzer could fire HE, AP, WP, and chemical smoke rounds. Maximum range of the weapon was 16,374 yards (14,972 meters). The M41 shared the same chassis and drive train with the M37, consequently the engines and performance characteristics were identical. The M41 operated with a five-man crew. The Massey Harris Company produced 60 M41s before World War Two ended in September of 1945.

Although the M37 and M41 HMCs were placed in production before the end of World War Two, neither vehicle saw combat during that conflict. The bulk of M37 and M41 production was completed after hostilities had ended. Both self-propelled guns were used in the Korean War and rendered invaluable fire support service. The M37 was also supplied under defense agreements to a number of American allies. These allies operated the M37 into the early 1970s.

The M41 remained in US Army service until the mid-1950s. The M37 served with Reserve and National Guard units until it was replaced by more modern equipment in the early 1960s. The M37 and M41 HMCs provided the bridge between the self-propelled artillery pieces from World War Two and the new weapons which resulted from wartime developments.

*For a more detailed look at these post-war weapons see M24 Chaffee in Action, Number 25, from Squadron/Signal Publications.

The M37 Howitzer Motor Carriage mounted a 105mm M4 howitzer on the new M24 Chaffee light tank chassis. This vehicle featured a similar arrangement to that of the earlier M7 HMC. Although produced too late to see service in World War Two, the M37 later saw combat during the Korean War. The M37 was also provided to a number of European nations under military aid programs. (PAM)

The M41 HMC mounted a 155mm M1 howitzer on the M24 chassis. This vehicle featured a similar arrangement to that of the earlier M12 HMC. The howitzer was mounted on the rear of the vehicle, which forced the engine to be relocated to the center of the chassis. The recoil spade is in the retracted position. The M41, like the M37, was produced too late to see combat in World War Two, however, it did see action in Korea. (PAM)

Tanks and Tank Killers from Squadron/Signal Publications

2033 M3 Lee/Grant

2034 M3 Half-track

2035 DUKW

2036 US Tank Destroyers

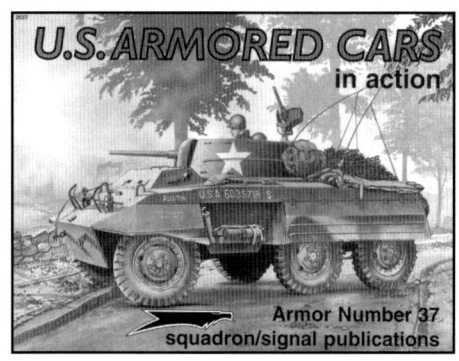

2037 US Armored Cars

1072 Hurricane

1073 Ju 87 Stuka

5511 P-47 Thunderbolt

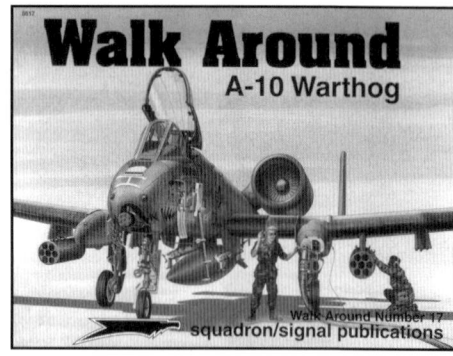

5517 A-10 Warthog